Contact the author:

- humbertcole3@gmail.com

- facebook.com/HumbertColeMath

Contents

Chapter 1

Introduction to algebra

What is algebra?

Algebra is the part of mathematics in which letters and other general symbols are used to represent numbers and quantities in expressions, equations and formulae.

Open Sentences

Example

$$14 + \boxed{} = 17$$

What number would enter the box to make the above statement true?
If you check it will be true if 3 goes into the box.

$$\text{i.e} \quad 14 + 3 = 17$$

Now try the next example

Example

$$4 - \boxed{} = 2$$

What number would enter the box to make the above statement true?
The answer is 2

$$\text{i.e} \quad 4 - 2 = 2$$

Now work through the following exercise in the same manner.

Exercises 1

In each sentence, find the number that makes the statement true. Use your knowledge of numbers.

1. $3 + 2 = \boxed{}$

2. $7 + 7 = \boxed{}$

3. $0 + 8 = \boxed{}$

4. $9 + 5 = \boxed{}$

5. $6 + 8 = \boxed{}$

6. $8 - 7 = \boxed{}$

7. $7 - 7 = \boxed{}$

8. $4 - 0 = \boxed{}$

9. $20 - 14 = \boxed{}$ 10. $11 - 11 = \boxed{}$

Letters For Numbers

- In mathematics, instead of using boxes as we did earlier we use letters of the alphabet to represent numbers

- We can write $5 + \boxed{}$ as $5 + x$

- Any letter can be used. For example, we may write $5 + a$, $5 + b$ or even $5 + c$

- When using a letter to stand for a number, the letter can stand for any number in general

- Thus the value of $5 + x$ depends on x

- If $x = 1$, $5 + x = 5 + 1 = 6$

- If $x = 2$, $5 + x = 5 + 2 = 7$

- If $x = 5$, $5 + x = 5 + 5 = 10$

- When letters and numbers are used in this way the mathematics is called **algebra**

- The statement $5 + x = 7$ is called an **algebraic sentence**. It means 5 plus a number x makes 7

Exercises 2

Find the number that each letter represents using your knowledge of numbers.

1. $5 + 2 = a$ 6. $7 - 3 = c$

2. $b = 2 + 2$ 7. $20 - 11 = y$

3. $2 + m = 5$ 8. $z = 5 - 5$

4. $x + 3 = 6$ 9. $20 - q = 16$

5. $22 + d = 30$ 10. $17 - f = 11$

Example   Find the number that each letter represents

(i) $8 \times a = 16$

(ii) $20 + b = 24$

Solution

(i) $8 \times a = 16$
   The number is 2. Therefore $a = 2$ because $8 \times 2 = 16$

(ii) $20 + b = 24$
   The number is 4. Therefore $b = 4$ because $20 + 4 = 24$

Example Find the number which each letter represents

1. $b + b = 18$

2. $x \times x \times x = 8$

Solution

1. $b + b = 18$
   The number is 9 because $9 + 9 = 18$. Therefore $b = 9$

2. $x \times x \times x = 8$
   The number is 2 because $2 \times 2 \times 2 = 8$ Therefore $x = 2$

Substitution in Algebraic Sentence

What is the value of $x + 6$?
The value of $x + 6$ depends on what x stands for. If $x = 3$, then $x + 6 = 3 + 6 = 9$. If $x = 8$, then $x + 6 = 8 + 6 = 14$.
Work through the following exercises using the same method.

Exercises 3

Find the value of the following when $x = 4$

1. $x + 5$

2. $x - 4$

3. $10 - x$

4. $9 + x$

5. $x - 2 + x$

6. $6 \times x$

7. $(x \times x) - 7$

8. $(x \div x) + x$

9. $16 \div x$

10. $(10 - x) \div 2$

Coefficients

- In arithmetic 3×4 is equal to $4 + 4 + 4$

$$\text{i.e} \quad 3 \times 4 = 4 + 4 + 4$$

- Similarly in algebra, $3 \times a$ is equal to $a + a + a$

$$\text{i.e} \quad 3 \times a = a + a + a$$

- Other examples are:

   1. $3 \times x = x + x + x$
   2. $4 \times z = z + z + z + z$
   3. $5 \times m = m + m + m + m + m$

- $3 \times a$ can be written as $3a$

  Hence,

  $$3 \times x = 3x$$
  $$4 \times z = 4z$$
  $$5 \times m = 5m$$

- Remember in arithmetic 3×2 is the same as 2×3 which is equal to 6.

  $$\text{i.e,} \quad 3 \times 2 = 2 \times 3 = 6$$

- Similarly in algebra $a \times b$ is the same as $b \times a$

- Examples are

  1. $x \times y = y \times x$
  2. $2 \times a = a \times 2$
  3. $2a \times b = b \times 2a$

Exercises 4

Simplify the following

1. $a + a$
2. $p + p$
3. $x + x + x + x$
4. $z + z + z + z + z$
5. $k + k + k + k + k$
6. $c + c + c + c + c$
7. $j + j$
8. $n + n + n + n$
9. $t + t + t + t$
10. $s + s + s + s + s$

Exercises 5

Expand the following. For example, $4x = x + x + x + x$

1. $3a$
2. $3m$
3. $7r$
4. $5x$
5. $6d$
6. $5s$

What is a coefficient?

- The coefficient of a letter in a term is simply the number multiplying the letter

- For example in the term $3x$, the coefficient of x is 3

- Other examples are

1. $10a$, the coefficient of a is 10
2. $3b$, the coefficient of b is 3
3. $7d$, the coefficient of d is 7

- Coefficient may be a whole number or a fraction. It may be positive or negative.

  1. $-4d$, the coefficient of d is -4
  2. $-10c$, the coefficient of c is -10
  3. $\frac{3}{7}b$, the coefficient of b is $\frac{3}{7}$
  4. $-\frac{2}{5}x$, the coefficient of x is $-\frac{2}{5}$

Exercises 6

What is the coefficient of each letter in the following terms.

1. $5a$
2. $-10d$
3. $17g$
4. $25n$
5. $-9m$

6. $\frac{2}{3}p$
7. $-\frac{5}{3}k$
8. $\frac{11}{9}z$
9. $\frac{y}{9}$
10. $\frac{9}{7}a$

Algebraic Expressions

- An algebraic expression consists of either a single term or a number of terms separated by $+$ and/or $-$ sign(s)

- $2x + 3y$ is an algebraic expression which consists of two terms: $2x$ and $3y$

- $8a$ is an algebraic expression which consists of only one term: $8a$

- $x - 2y$ is an algebraic expression which consists of two terms: x and $-2y$

- $2x + 3y - 8a$ is an algebraic expression which consists of three terms: $2x$, $3y$ and $-8a$

- A number may also be a term in an algebraic expression. Such a number is called a **numerical term** or a **number term**

- $3 + 6a$ is an algebraic expression with two terms: 3 and $6a$

Exercises 7

Write the term(s) in each of the following expressions

1. $2a + 3b$

2. $5k - h$

3. $6f - 2g + 3k + 5$

4. $6 - 8a$

5. $-14s - 8t + 22r$

6. $\frac{5}{8}x$

7. $\frac{3}{11}y + 7$

8. $\frac{5}{6}x - \frac{10}{3}y$

9. $\frac{7}{3}h - 4$

10. $3 - 10x$

Addition and Subtraction Operations on Like Terms

Like Terms: Like terms are a group of terms which contain the same letter. For example $2x$, x and $\frac{1}{3}x$ are like terms because they all contain the same letter x.

Unlike Terms: Unlike terms are a group of terms which contain different letters. For example $2a$, $3x$ and $5c$ are unlike terms because they contain the different letters.

Example Identify which of the following groups of terms are like terms.

1. $2a, 5a, 6a$

2. $2a, b, d$

3. b, c, d

4. a, b, c

5. $x, -3x, \frac{3}{8}x$

6. $7z, 80z, \frac{z}{8}$

Answers

1. like terms

2. unlike terms

3. unlike terms

4. unlike terms

5. like terms

6. like terms

Exercises 8

Which of the following groups of terms are like terms? Which of them are unlike terms?

1. $x, 5x, 3x, 10x$

2. $p, 2p, 5p$

3. m, n, o

4. $2t, -3t, -\frac{5}{3}t$

5. $k, -\frac{2}{3}k$

6. $\frac{11}{3}x, -\frac{5}{3}t$

7. $\frac{5}{3}m, \frac{5}{3}n$

8. $-120k, -120p$

9. $16x, 16y$

10. $j, -3j, 5j$

Addition Operation

Addition of two terms will yield a simpler result only if the terms are like terms

$$\text{e.g.} \quad x + 3x = (x) + (x + x + x)$$

$$= x + x + x + x$$
$$= 4x$$

If the terms are unlike terms there would be no simpler result

e.g. $\quad y + 3x \quad$ cannot be further simplified

Example Perform the addition operation in the following expressions. If the expression cannot be further simplified, say so.

1. $a + a$

2. $a + 2a$

3. $3x + 5x$

4. $5d + 4k$

Solution

1. $a + a = 2a$

2. $a + 2a = a + (a + a) = a + a + a = 3a$

3.

$$3x + 5x = (x + x + x) + (x + x + x + x + x)$$
$$= x + x + x + x + x + x + x + x = 8x$$

4. $5d + 4k$
   This expression cannot be further simplified because $5d$ and $4k$ are **unlike terms**

Shortcut For Addition Operation

If for example you are given an expression $10x + 2x$ and you are asked to simplify, it will be very tiresome if you evaluate it using the method explained previously.
If we go by the previous method we will have

$$10x + 2x = (x + x + x + x + x + x + x + x + x + x) + (x + x)$$
$$= x + x + x + x + x + x + x + x + x + x + x + x$$
$$= 12x$$

* In the shortcut method to simplify $10x + 2x$ simply follow these steps

Step 1: Ignore x and add like in arithmetic

i.e. $\quad 10x + 2x \xrightarrow{\text{ignore } x} 10 + 2$
$$10 + 2 = 12$$

Step 2:   Restore x to the result

$$12 \xrightarrow{\text{restore } x} 12x$$

therefore, the answer is $12x$

Example   Simplify the following expressions. If further simplification is not possible, say so. Use the shortcut method.

1. $20a + 53a$

2. $x + 15x + 6x$

3. $15y + 6y + 7y + y$

Solution

1. $20a + 53a$

    Step 1:   Ignore a and add arithmetically

    $$20a + 53a \xrightarrow{\text{ignore } a} 20 + 53$$
    $$20 + 53 = 73$$

    Step 2:   Restore a

    $$73 \xrightarrow{\text{restore } a} 73a$$

    therefore, the answer is $73a$

2. $x + 15x + 6x$

    Step 1:   Ignore x and add arithmetically

    $$x + 15x + 6x \xrightarrow{\text{ignore } x} 1 + 15 + 6$$
    $$1 + 15 + 6 = 22$$

    Step 2:   Restore x

    $$22 \xrightarrow{\text{restore } x} 22x$$

    therefore, the answer is $22x$

3. $15y + 6y + 7y + y$

    Step 1:   Ignore y and add arithmetically

    $$15y + 6y + 7y + y \xrightarrow{\text{ignore } y} 15 + 6 + 7 + 1$$
    $$15 + 6 + 7 + 1 = 29$$

Step 2: Restore y

$$29 \xrightarrow{\text{restore } y} 29y$$

therefore, the answer is $29y$

Exercises 9

Simplify the following where possible

1. $z + 10z + 2z$

2. $11m + 12n$

3. $16p + 2p + p$

4. $p + p + 3p$

5. $7a + a + 8a + a$

6. $h + 11h$

7. $7d + 5d + d$

8. $k + 5d$

9. $5x + 2x + x$

10. $x + t + h$

Subtraction Operation

Example Simplify the following

1. $5x - 3x$

2. $10y - 3y$

3. $6a - 3a - 2a$

Solution

1. $5x - 3x$

 Step 1: Ignore x and subtract arithmetically

$$5x - 3x \xrightarrow{\text{ignore } x} 5 - 3$$
$$5 - 3 = 2$$

 Step 2: Restore x

$$2 \xrightarrow{\text{restore } x} 2x$$

 therefore, the answer is $2x$

2. $10y - 3y$

 Step 1: Ignore y and subtract arithmetically

$$10y - 3y \xrightarrow{\text{ignore } y} 10 - 3$$
$$10 - 3 = 7$$

Step 2:   Restore y

$$7 \xrightarrow{\text{restore } y} 7y$$

therefore, the answer is $7y$

3. $6a - 3a - 2a$

Step 1:   Ignore a and subtract arithmetically

$$6a - 3a - 2a \xrightarrow{\text{ignore } a} 6 - 3 - 2$$
$$6 - 3 - 2 = 3 - 2 = 1$$

Step 2:   Restore a

$$1 \xrightarrow{\text{restore } a} 1a$$
$$1a = a$$

therefore, the answer is a

Exercises 10

Simplify the following where possible

1. $6x - 4x$

2. $8m - 5n$

3. $a - 3b$

4. $8y - 3y - 4y$

5. $2b - b - b$

6. $5a - 3a$

7. $200a - 5a - 100a$

8. $16a - 5x$

9. $35p - 10p$

10. $60z - 30z - 19z - z$

Combined Addition and Subtraction Operation

Example   Simplify the following where possible

1. $5a + 2a - 3a + a$

2. $100x - 2x - 15x + 6x - 20x$

3. $120y - 119y + y$

Solution

1. $5a + 2a - 3a + a$

Step 1: Ignore a and perform addition and subtraction arithmetically

$$5a + 2a - 3a + a \xrightarrow{\text{ignore } a} 5 + 2 - 3 + 1$$
$$5 + 2 - 3 + 1 = 7 - 3 + 1 = 4 + 1 = 5$$

Step 2: Restore a

$$5 \xrightarrow{\text{restore } a} 5a$$

therefore, the answer is $5a$

2. $100x - 2x - 15x + 6x - 20x$

Step 1: Ignore x and perform addition and subtraction arithmetically

$$100x - 2x - 15x + 6x - 20x \xrightarrow{\text{ignore } x} 100 - 2 - 15 + 6 - 20$$
$$= 98 - 15 + 6 - 20$$
$$= 83 + 6 - 20$$
$$= 89 - 20$$
$$= 69$$

Step 2: Restore x

$$69 \xrightarrow{\text{restore } x} 69x$$

(This is the final answer)

3. $120y - 119y + y$

Step 1: Ignore y and perform addition and subtraction arithmetically

$$120y - 119y + y \xrightarrow{\text{ignore } y} 120 - 119 + 1$$
$$120 - 119 + 1 = 1 + 1 = 2$$

Step 2: Restore y

$$2 \xrightarrow{\text{restore } y} 2y$$

Exercises 11

Simplify the following where possible.

1. $2a - a + 2a$

2. $5x + 10x - 11x$

3. $z - z + 10z - 9z$

4. $16g - 9g - g - 3g$

5. $j + 2j - j$

6. $22e - 5e - 8e - 4e$

7. $12f - 5f - 3f - 4f$

8. $a - b - c - d$

9. $6g + 5g - 2g - 4g - 2g$

10. $2q - 20q + 11q + 8q + 5q$

Grouping Like and Unlike Terms

- Consider the algebraic expression below

$$8x + 2y - 4x - 3x + 9y$$

- This expression can be further simplified

- In order to simplify this expression we must group all like terms together

- That means

$$8x + 2y - 4x - 2x + 9y \qquad \text{(before grouping)}$$
$$= 8x - 4x - 2x + 2y + 9y \qquad \text{(after grouping)}$$

- Now we will evaluate each group separately

(i) $8x - 4x - 2x = 4x - 2x = 2x$

(ii) $2y + 9y = 11y$

Therefore

$$8x - 4x - 2x + 2y + 9y$$
$$= 2x + 11y$$

Example   Simplify the following where possible

1. $2x + 13 + 10 + 5x - 16$

2. $10k + 13g + 3k - 8g - 6k$

3. $x + y + 2x + 3y$

4. $10 + 2p + 21 - 3 + 5p$

Solution

1. $2x + 13 + 10 + 5x - 16$

Step 1: Group like terms together

$$2x + 13 + 10 + 5x - 16 \qquad \text{(before grouping)}$$
$$= 2x + 5x + 13 + 10 - 16 \qquad \text{(after grouping)}$$

Step 2: Evaluate each group separately

(a) $2x + 5x = 7x$

(b) $13 + 10 - 16 = 9$

Therefore

$$2x + 5x + 13 + 10 - 16 = 7x + 9$$

Therefore, the final answer is $7x + 9$

2. $10k + 13g + 3k - 8g - 6k$

Step 1: Group like terms together

$$10k + 13g + 3k - 8g - 6k \qquad \text{(before grouping)}$$
$$= 10k + 3k - 6k + 13g - 8g \qquad \text{(after grouping)}$$

Step 2: Evaluate each group separately and add the results together

(a) $10k + 3k - 6k = 13k - 6k = 7k$

(b) $13g - 8g = 5g$

Therefore

$$10k + 3k - 6k + 13g - 8g = 7k + 5g$$

3. $x + y + 2x + 3y$

Step 1: Group like terms together

$$x + y + 2x + 3y \qquad \text{(before grouping)}$$
$$= x + 2x + y + 3y \qquad \text{(after grouping)}$$

Step 2: Evaluate each group separately and add the results together

(a) $x + 2x = 3x$

(b) $y + 3y = 4y$

Therefore, answer is $3x + 4y$

4. $10 + 2p + 21 - 3 + 5p$

$$10 + 2p + 21 - 3 + 5p \qquad \text{(before grouping)}$$
$$= 10 + 21 - 3 + 2p + 5p \qquad \text{(after grouping)}$$

(a) $10 + 2 - 3 = 9$

(b) $2p + 5p = 7p$

Add the results together

$$9 + 7p$$

The answer is $9 + 7p$
(Note that $9 + 7p$ is the same as $7p + 9$)

Exercises 12

Group like terms and simplify the following expressions, where possible.

1. $8c + 18d + 30c - 8d$

2. $4y - 2x + 5x - 3y$

3. $6c - 10c + 13c - 7$

4. $12n - 3p - 3p - 5p$

5. $2x - 8 - 3 + 5x$

6. $2a - 5a + 14 - 6$

7. $2x + 3x + 7$

8. $6a + 2a + 3 + 10$

9. $3x + 9y + 8x - 4y$

10. $8a - 5a + 14 - 20$

11. $6p - 2q + 4r - 2p + 3s + 5q$

12. $2a + b + c + d + 5e$

13. $6m + 3n - 1 - 6m + 4$

14. $5h + 8k + 2 - 3k - 2h$

15. $a - 7b + 3c + 8b + 2a$

16. $3x + 2 - 7x - 4 + 5x + 6$

17. $7m - 2n + 6 - 5m + 7n + 3$

18. $5f - 3g - 7f + 9g + 3f - 4g$

19. $13a + b - 9a + 7b$

20. $6m - 9m + 4m$

Solution of Exercises

Exercises 1

1. 5

2. 14

3. 8

4. 14

5. 14

6. 1

7. 0

8. 4

9. 6

10. 0

Exercises 2

1. $a = 7$
2. $b = 4$
3. $m = 3$
4. $x = 3$
5. $d = 8$

6. $c = 4$
7. $y = 9$
8. $z = 0$
9. $q = 4$
10. $f = 6$

Exercises 3

1. 9
2. 0
3. 6
4. 13
5. 6

6. 24
7. 9
8. 5
9. 4
10. 3

Exercises 4

1. $2a$
2. $2p$
3. $4x$
4. $5z$
5. $5k$

6. $5c$
7. $2j$
8. $4n$
9. $4j$
10. $5s$

Exercises 5

1. $a + a + a$
2. $m + m + m$
3. $r + r + r + r + r + r + r$

4. $x + x + x + x + x$
5. $d + d + d + d + d + d$
6. $s + s + s + s + s$

Exercises 6

1. 5
2. -10
3. 17
4. 25

5. -9
6. $\frac{2}{3}$
7. $-\frac{5}{3}$
8. $\frac{11}{9}$

9. $\frac{1}{9}$

10. $\frac{9}{7}$

Exercises 7

1. $2a$ and $3b$

2. $5k$ and $-h$

3. $6f$, $-2g$, $3k$ and 5

4. 6 and $-8a$

5. $-14s$, $-8t$ and $22r$

6. $\frac{5}{8}x$

7. $\frac{3}{11}y$ and 7

8. $\frac{5}{6}x$ and $\frac{10}{3}y$

9. $\frac{7}{3}h$ and -4

10. 3 and $-10x$

Exercises 8

1. like terms

2. like terms

3. unlike terms

4. like terms

5. like terms

6. unlike terms

7. unlike terms

8. unlike terms

9. unlike terms

10. like terms

Exercises 9

1. $13z$

2. $11m + 12n$

3. $19p$

4. $5p$

5. $17a$

6. $12h$

7. $13d$

8. $k + 5d$

9. $8x$

10. $x + t + h$

Exercises 10

1. $2x$

2. $8m - 5n$

3. $a - 3b$

4. y

5. 0

6. $2a$

7. $95a$

8. $16a - 5x$

9. $25p$

10. $10z$

Exercises 11

1. $3a$

2. $6x$

3. z

4. $3g$

5. $2j$

6. $5e$

7. 0

8. $a - b - c - d$

9. $3g$

10. $6q$

Exercises 12

1. $38c - 10d$

2. $y + 3x$

3. $9c - 7$

4. $12n - 11p$

5. $7x - 11$

6. $8 - 3a$

7. $5x + 7$

8. $8a + 13$

9. $11x + 5y$

10. $3a - 6$

11. $4p + 3q + 4r + 3s$

12. $2a + b + c + d + 5e$

13. $3n + 3$

14. $3h + 5k + 2$

15. $3a + b + 3c$

16. $x + 4$

17. $5n + 2m + 9$

18. $f + 2g$

19. $4a + 8b$

20. m

Chapter 2

Algebraic expressions

Multiplying Algebraic Terms

One of the ways we represent multiplication in algebra is concatenation. For example, we represent $a \times b$ by ab.

FACT 1: Just as $2 \times a$ is equal to $2a$ so also $a \times b$ is equal to ab.
Other examples are:

1. $x \times y = xy$

2. $a \times b = ab$

3. $x \times y \times z = xyz$

4. $10 \times p \times q = 10pq$

5. $10 \times p \times q \times r = 10pqr$

FACT 2: Just as $5 \times 5 = 5^2$ and $5 \times 5 \times 5 = 5^3$ so also $a \times a = a^2$ and $a \times a \times a = a^3$
Other examples are:

1. $x \times x = x^2$

2. $x \times x \times x = x^3$

3. $b \times b = b^2$

4. $c \times c \times c \times c = c^4$

5. $y \times y \times y \times y = y4$

Exercises 1

Simplify the following:

1. $r \times k$

2. $h \times b$

3. $c \times b$

4. $j \times k \times l$

5. $m \times n$

6. $m \times n \times p$

7. $a \times b \times c \times d$

8. $b \times b \times b$

9. $k \times k \times k \times k \times k \times k$

10. $m \times m \times m$

FACT 3:   Just as 2×3 is equal to 3×2 so also $a \times b$ is equal to $b \times a$

$$\text{i.e} \quad ab = ba$$

Other examples are:

1. $x \times y = y \times x$

$$xy = yx$$

2. $mn = nm$

3. $xz = zx$

4. $xyz = zxy = yzx$, etc

5. $hkl = hlk = khl$, etc

FACT 4:   Just as $8 \times 6 \times 6 = 8 \times 6^2$
so also $a \times b \times b = a \times b^2 = ab^2$
Other examples are:

1. $a \times a \times b = a^2 \times b = a^2 b$

2. $2 \times b \times b = 2 \times b^2 = 2b^2$

3. $8 \times x \times x \times x = 8 \times x^3 = 8x^3$

4. $a \times a \times b \times b = a^2 \times b^2 = a^2 b^2$

5. $x \times x \times x \times y \times y \times y = x^3 \times y^3 = x^3 y^3$

Exercises 2

Simplify the following

1. $x \times x \times a$

2. $5 \times a \times a$

3. $b \times b \times a$

4. $a \times a \times t$

5. $m \times m \times x$

6. $16 \times y \times z \times z$

7. $x \times x \times y \times y \times z \times z$

8. $p \times p \times z$

9. $a \times p \times p \times p$

10. $22 \times a \times a \times a \times b \times b \times b$

FACT 5:   Remember again that order is not important in algebraic terms. Examples are:

1. $a^2b = ba^2$

   (Why? because $a^2b = a^2 \times b = b \times a^2 = ba^2$)

2. $8a^2b^2 = 8b^2a^2$

3. $x^3y^3 = y^3x^3$

4. $b^2z^2 = z^2b^2$

5. $4x^2b^2 = 4b^2x^2$

The only rule to remember about ordering is that in a term the number must be written before the letters.

That means we write

$$b \times 8 \quad \text{as} \quad 8b$$

$$\text{and not as } b8$$

$$b \times 8 = 8b \quad \text{(correct)}$$
$$b \times 8 = b8 \quad \text{(wrong)}$$

FACT 6: Grouping Like Letters   When asked to simplify $5 \times 2 \times 5 \times 2 \times 5$, you can rearrange the numbers and group all similar numbers together.

That is,

$$5 \times 2 \times 5 \times 2 \times 5$$

$\rightarrow$ take all the 5's to one side and all the 2's to the other side

$$= 5 \times 5 \times 5 \times 2 \times 2$$
$$= 5^3 \times 2^2$$

So also in algebra when asked to simplify $a \times b \times a \times a \times b$, you can rearrange the letters and group similar letters together. Take all the a's to one side and take all the b's to the other side.

That is,

$$a \times b \times a \times a \times b = a \times a \times a \times b \times b$$
$$= a^3 \times b^2$$

Example   Simplify the following

1. $s \times t \times s \times t \times t$

2. $2 \times x \times y \times 3 \times x \times 4 \times y$

Solution

1. $s \times t \times s \times t \times t$

   Take all the s's to one side and all the t's to the other

   $$= s \times s \times t \times t \times t$$
   $$= s^2 \times t^3$$

2. $2 \times x \times y \times 3 \times x \times 4 \times y$

   Take all the numbers to one side, all the y's to one side and all the x's to one side.

   $$= 2 \times 3 \times 4 \times x \times x \times y \times y$$
   $$= 24 \times x^2 \times y^2$$
   $$= 24x^2y^2 \quad \longleftarrow \text{ Final Answer}$$

   You could have also put the y letters in the middle, such that

   $$= 2 \times 3 \times 4 \times y \times y \times x \times x$$
   $$= 24 \times y^2 \times x^2$$
   $$= 24y^2x^2 \quad \longleftarrow \text{ Final Answer}$$

   Note that $24y^2x^2$ and $24x^2y^2$ are the same and are both correct.

Exercises 3

Simplify the following:

1. $a \times 3 \times a \times 2$

2. $p \times p \times q \times p$

3. $5 \times 4 \times k \times 2$

4. $t \times m \times n \times m$

5. $g \times h \times 3 \times h$

6. $11 \times s \times 5 \times t$

7. $a \times b \times c \times c \times a \times b$

8. $2 \times b \times c \times 5 \times a \times b$

9. $100 \times 2b \times 6a \times 15b \times b$

10. $5x \times y \times 11 \times x \times y$

FACT 7:

Laws of Indices

According to the laws of indices

$$2^2 \times 2^5 = 2^{2+5} = 2^7$$

So also in algebra

$$x^2 \times x^5 = x^{2+5} = x^7$$

Other examples are

1. $b \times b^3 = b^{1+3} = b^4$

2. $a^4 \times a^2 \times a = a^{4+2+1} = a^{6+1} = a^7$

3. $y^2 \times y^3 = y^{2+3} = y^5$

Exercises 4

Simplify the following using the laws of indices

1. $a^4 \times a^5$

2. $b^6 \times b^4$

3. $a \times a \times a^6$

4. $c \times c^5$

5. $z^2 \times z$

6. $a \times a^6 \times b^2 \times b^3$

7. $m^2 \times n^2 \times m^2 \times n$

8. $5 \times a^2 \times a^2$

9. $2 \times b^2 \times b^3$

10. $10 \times y^2 \times y^3 \times 3$

Division of Algebraic Terms

Dividing an algebraic term by a number.

Example   Simplify the following

1. $4x \div 2$

2. $10y \div 5$

3. $16z \div 8$

4. $100p^2 \div 5$

5. $12st \div 4$

6. $11xy^2 \div 11$

Solution   To solve all the problems listed above simply ignore the algebraic part then divide arithmetically and finally restore the arithmetic part of the result.

1. $4x \div 2$

Step 1: ignore x

$$4x \div 2 = \frac{4x}{2}$$

$$\frac{4x}{2} \xrightarrow{\text{ignore } x} \frac{4}{2}$$

$$\frac{4}{2} = 2$$

Step 2: restore x

$$2 \xrightarrow{\text{restore } x} 2x$$

$\therefore$ the answer is $2x$

2. $10y \div 5$

Step 1: ignore y

$$10y \div 5 = \frac{10y}{5}$$
$$\frac{10y}{5} \xrightarrow{\text{ignore } y} \frac{10}{5}$$
$$\frac{10}{5} = 2$$

Step 2: restore y

$$2 \xrightarrow{\text{restore } y} 2y$$

$\therefore$ the answer is $2y$

3. $16z \div 8$

Step 1: ignore z

$$\frac{16z}{8} \xrightarrow{\text{ignore } z} \frac{16}{8}$$
$$\frac{16}{8} = 2$$

Step 2: restore z

$$2 \xrightarrow{\text{restore } z} 2z$$

$\therefore$ the answer is $2z$

4. $100p^2 \div 5$

Step 1: ignore p^2

$$\frac{100p^2}{5} \xrightarrow{\text{ignore } p^2} \frac{100}{5}$$
$$\frac{100}{5} = 20$$

Step 2: restore p^2

$$20 \xrightarrow{\text{restore } p^2} 20p^2$$

$\therefore$ the answer is $20p^2$

5. $12st \div 4$

Step 1: ignore st

$$\frac{12st}{4} \xrightarrow{\text{ignore } st} \frac{12}{4}$$
$$\frac{12}{4} = 3$$

Step 2: restore st

$$3 \xrightarrow{\text{restore } st} 3st$$

$\therefore$ the answer is $3st$

6. $11xy^2 \div 11$

Step 1: ignore xy^2

$$\frac{11xy^2}{11} \xrightarrow{\text{ignore } xy^2} \frac{11}{11}$$
$$\frac{11}{11} = 1$$

Step 2: restore xy^2

$$1 \xrightarrow{\text{restore } xy^2} 1xy^2$$
$$1xy^2 = xy^2$$

$\therefore$ the answer is xy^2

Exercises 5

Simplify the following where possible

1. $8a \div 2$

2. $\frac{15a}{3}$

3. $\frac{6x}{2}$

4. $\frac{16x}{5}$

5. $\frac{130b}{2}$

6. $\frac{2000g}{5}$

7. $\frac{9x^2y^2}{3}$

8. $\frac{25xyz}{5}$

9. $10xt \div 2$

10. $12de \div 5$

Dividing an Algebraic Term by Another Algebraic Term

Just as $\frac{8}{8} = 1$ in arithmetic so also in algebra $\frac{a}{a} = 1$
Other examples are:

1. $\frac{x}{x} = 1$

2. $\frac{xy}{xy} = 1$

3. $\frac{abc}{abc} = 1$

4. $\frac{x^2}{x^2} = 1$

5. $\frac{2a}{2a} = 1$

6. $\frac{6x^2y}{6x^2y} = 1$

Thus generally if the numerator and denominator are the same the result of the division operation is always 1.

Division by Cancellation

Evaluate $\frac{6\times3}{3}$

The result of $\frac{6\times3}{3}$ is 6

Because $\frac{6\times\cancel{3}}{\cancel{3}} = 6$

When we cancel a number or letter in the numerator in cancellation method of division, it means the letter has an equivalent in the denominator.

Example Evaluate $\frac{10\times2\times10}{10\times5\times2}$

1. Cancel any pair of similar letters that appear at numerator (top) and denominator (bottom)

$\rightarrow \frac{\cancel{10}\times\cancel{2}\times10}{\cancel{10}\times5\times\cancel{2}}$

The pairs found were pairs of 10 and 2

2. Write down what is left and evaluate

$$\rightarrow \frac{10}{5}$$

What is left is $\frac{10}{5}$

$\frac{10}{5} = 2$

$\therefore$ The answer is 2

Applying Cancellation Method in Algebraic Division

Example Simplify the following using cancellation method

1. $\frac{x\times x\times y}{z\times x\times y}$

2. $\frac{a\times a\times b}{b\times c\times d\times a}$

3. $\frac{h\times h\times h}{a\times h\times h}$

Solution

1. $\frac{x \times x \times y}{z \times x \times y}$

   (a) Cancel pair of similar letters

   $$\rightarrow \frac{x \times \cancel{x} \times \cancel{y}}{z \times \cancel{x} \times \cancel{y}}$$

   (b) Write down what is left

   $$= \frac{x}{z}$$

   $\therefore$ the answer is $\frac{x}{z}$

2. $\frac{a \times a \times b}{b \times c \times d \times a}$

   (a) Cancel pair of similar letters

   $$\rightarrow \frac{\cancel{a} \times a \times \cancel{b}}{\cancel{b} \times c \times d \times \cancel{a}}$$

   (b) Write down what is left

   $$= \frac{a}{c \times d} = \frac{a}{cd}$$

   $\therefore$ the answer is $\frac{a}{cd}$

3. $\frac{h \times h \times h}{a \times h \times h}$

   (a) Cancel pair of similar letters

   $$\rightarrow \frac{h \times \cancel{h} \times \cancel{h}}{a \times \cancel{h} \times \cancel{h}}$$

   (b) Write down what is left

   $$= \frac{h}{a}$$

   $\therefore$ the answer is $\frac{h}{a}$

Example   Simplify the following using the cancellation method.

1. $\frac{x^2 y}{zxy}$

2. $\frac{a^2 b}{bcda}$

3. $\frac{m^2 n^3}{mn}$

4. $\frac{8x^2}{x}$

5. $\frac{10y^2 x}{5x^2 y}$

6. $\frac{16pst}{11s^2 p}$

Solution

1. $\frac{x^2y}{zxy}$

   (a) Expand all the terms then cancel pair of similar letters

   $$\rightarrow \frac{x^2y}{zxy} = \frac{x \times x \times y}{z \times x \times y} = \frac{x \times \cancel{x} \times \cancel{y}}{z \times \cancel{x} \times \cancel{y}}$$

   (b) Write down what is left

   $$\rightarrow \frac{x}{z}$$

   $\therefore$ the answer is $\frac{x}{z}$

2. $\frac{a^2b}{bcda}$

   (a) Expand all terms and cancel pair of similar letters

   $$\rightarrow \frac{a^2b}{bcda} = \frac{a \times a \times b}{b \times c \times d \times a} = \frac{a \times \cancel{a} \times \cancel{b}}{\cancel{b} \times c \times d \times \cancel{a}}$$

   (b) Write down what is left

   $$\rightarrow \frac{a}{c \times d} = \frac{a}{cd}$$

   $\therefore$ the answer is $\frac{a}{cd}$

3. $\frac{m^2n^3}{mn}$

   (a) Expand all terms and cancel pairs of similar letters

   $$\frac{m^2n^3}{mn} = \frac{m \times m \times n \times n \times n}{m \times n} = \frac{m \times \cancel{m} \times n \times n \times \cancel{n}}{\cancel{m} \times \cancel{n}}$$

   (b) Write down what is left

   $$\rightarrow m \times n \times n = m \times n^2 = mn^2$$

   $\therefore$ the answer is mn^2

4. $\frac{8x^2}{x}$

   (a) Expand the terms and cancel pairs of similar letters

   $$\rightarrow \frac{8x^2}{x} = \frac{8 \times x \times x}{x} = \frac{8 \times x \times \cancel{x}}{\cancel{x}}$$

   (b) Write down what is left

   $$\rightarrow 8 \times x = 8x$$

5. $\frac{10y^2x}{5x^2y}$

   (a) Expand the terms and cancel pairs of similar letters

   $$\rightarrow \frac{10y^2x}{5x^2y} = \frac{10 \times y \times y \times x}{5 \times x \times x \times y} = \frac{10 \times y \times \cancel{y} \times \cancel{x}}{5 \times x \times \cancel{x} \times \cancel{y}}$$

(b) Write down what is left

$$\rightarrow \frac{10 \times y}{5 \times x} = \frac{10y}{5x} = \frac{2y}{x} \quad \left(\text{because } \frac{10}{5} = 2 \right)$$

$\therefore$ the answer is $\frac{2y}{x}$

6. $\frac{16pst}{11s^2p}$

(a) Expand the terms and cancel pairs of similar letters

$$\rightarrow \frac{16pst}{11s^2p} = \frac{16 \times p \times s \times t}{11 \times s \times s \times p} = \frac{16 \times \cancel{p} \times \cancel{s} \times t}{11 \times \cancel{s} \times s \times \cancel{p}}$$

(b) Write down what is left

$$\rightarrow \frac{16 \times t}{11 \times s} = \frac{16t}{11s}$$

$\therefore$ the answer is $\frac{16t}{11s}$

Exercises 6

Simplify the following using division by cancellation where possible.

1. $\frac{6a}{a}$

2. $\frac{18y}{6x}$

3. $\frac{1}{8}$ of $32x$

4. $28ab \div 4b^2$

5. $17mn \div n$

6. $d^2 \div d$

7. $3x^2 \div x$

8. $\frac{7x^3}{x}$

9. $\frac{12x^2}{3x}$

10. $\frac{33mn}{8a}$

11. $\frac{z^2}{2}$

12. $\frac{5c^2}{c}$

13. $\frac{5x^3}{x^2}$

14. $\frac{26b^4}{b}$

15. $\frac{54ab}{9b}$

16. $\frac{40pq^2}{pq}$

17. $\frac{48z^2y}{12zy}$

18. $\frac{72a^2b}{8a}$

19. $\frac{a^2b^2c^2}{abc}$

20. $15dq^2 \div 3q$

Division using the Laws of Indices

Laws of Indices: According to the laws of indices

$$\frac{2^5}{2^2} = 2^{5-2} = 2^3$$

So also in algebra

$$\frac{x^5}{x^2} = x^{5-2} = x^3$$

Other examples are:

1. $\frac{b^{10}}{b^6} = b^{10-6} = b^4$

2. $\frac{y^7}{y^6} = y^{7-6} = y^1 = y$

3. $\frac{z^{20}}{z^9} = z^{20-9} = z^{11}$

4. $\frac{a^2 b^3}{ab^2} = a^{2-1} \times b^{3-2} = ab^2$

According to the laws of indices

$$2^{-6} = \frac{1}{2^6} \quad \text{and so} \quad 2^{-1} = \frac{1}{2^1} = \frac{1}{2}$$

So also in algebra, $x^{-6} = \frac{1}{x^6}$ and so $x^{-1} = \frac{1}{x}$
Other examples are:

1. $b^{-10} = \frac{1}{b^{10}}$

2. $\frac{g^6}{g^{10}} = g^{6-10} = g^{-4} = \frac{1}{g^4}$

3. $\frac{h^7}{h^{13}} = h^{7-13} = h^{-6} = \frac{1}{h^6}$

4. $\frac{ab^3}{a^3 b^4} = a^{1-3} \times b^{3-4} = a^{-1} \times b^{-1} = \frac{1}{a} \times \frac{1}{b} = \frac{1}{ab}$

The division by cancellation method is used in simplifying algebraic expressions that have very high powers.
If for example you are asked to simplify the expression

$$\frac{a^{16} \times b^{10}}{a^5 \times b^7}.$$

It will be very stressful to expand the expression first and then simplify by cancellation.
It is easier to use the indices method. The indices method is best suited for problems such as this.

Example Simplify the following expressions using the indices method.

1. $\frac{y^7}{y^5}$

2. $\frac{a^{16} \times b^{10}}{a^5 \times b^7}$

3. $\frac{xy^2 z}{x^2 y^5 z^4}$

4. $\frac{15h^2}{3h^4 kl}$

5. $\frac{8m^2 n}{32mn}$

Solution

1. $\frac{y^7}{y^5}$

$$= y^{7-5} = y^2$$

2. $\frac{a^{16} \times b^{10}}{a^5 \times b^7}$

$$= a^{16-5} \times b^{10-7}$$
$$= a^{11} \times b^3$$
$$= a^{11}b^3$$

3. $\frac{xy^2z}{x^2y^5z^4}$

$$= x^{1-2} \times y^{2-5} \times z^{1-4}$$
$$= x^{-1} \times y^{-3} \times z^{-3}$$
$$= \frac{1}{x} \times \frac{1}{y^3} \times \frac{1}{z^3}$$
$$= \frac{1}{xy^3z^3}$$

4. $\frac{15h^2}{3h^4kl}$

$$= \frac{15}{3} \times h^{2-4} \times \frac{1}{k} \times \frac{1}{l}$$
$$= 5 \times h^{-2} \times \frac{1}{k} \times \frac{1}{l}$$
$$= 5 \times \frac{1}{h^2} \times \frac{1}{k} \times \frac{1}{l}$$
$$= \frac{5}{h^2kl}$$

5. $\frac{8m^2n}{32mn}$

$$= \frac{8}{32} \times \frac{m^2}{m} \times \frac{n}{n}$$
$$= \frac{1}{3} \times m^{2-1} \times n^{1-1}$$
$$= \frac{1}{3} \times m^1 \times n^0$$
$$= \frac{1}{3} \times m \times 1$$
$$= \frac{m}{3}$$

Exercises 7

Simplify the following

1. $24a^4b^4 \div 3a$

2. $8a^2 \div 4abc^4$

3. $\frac{xy^2z^{16}}{x^{16}y^2z}$

4. $\frac{y^{16}}{2y^8}$

5. $\frac{b^2x^2}{abx^5}$

6. $\frac{18a^{20}}{3a^2b}$

7. $\frac{16h^3}{h^5k}$

8. $\frac{1}{4x} \times 8x^5$

9. $\frac{8y}{y}$

10. $\frac{7a}{8a^{100}}$

Order of Operations

Consider the expression below:

$$17 - 5 \times 12$$

There are two possible answers depending on whether we carry out the subtraction first or whether we carry out the multiplication first.

Subtraction first:

$$(17 - 5) \times 12 = 12 \times 12 = 144$$

Multiplication first:

$$17 - (5 \times 12) = 17 - 60 = -43$$

In mathematics the law which guides the decision of which operation to carry out first is summarized as PEMDAS.

P - Parentheses
E - Exponents
M - Multiplication
D - Division
A - Addition
S - Subtraction

PEMDAS is a mnemonic to help you remember the correct order to complete mathematical operations.

How to Apply PEMDAS rule

1. **Parentheses:** Start with anything inside parentheses (brackets)

2. **Exponents:** After brackets, work on the terms having the powers or roots

3. **Multiplication and Division:** Multiplication and Division come after order. They have the same rank meaning you can choose to do anyone first.

4. **Addition and Subtraction:** Addition and Subtraction is the final step. Addition and subtraction have the same rank meaning you can choose to do any one first.

Very Important Note: When applying the rule of PEMDAS always work from left to right.

Example   Simplify the following

1. $3 + 20 \times 3$

2. $25 - 5 \div (3 + 2)$

3. $10 + 6 \times (1 \div 10)$

4. $5(3 + 2) + 5^2$

5. $(105 + 206) - 550 \div 5^2 + 10$

6. $7 + 7 \div 7 + 7 \times 7$

Solution

1. $3 + 20 \times 3$

   There are no brackets or orders in this sum

   (a) Multiplication comes before addition, so you can start with $20 \times 3 = 60$

   (b) Your sum now reads $3 + 60$
       The answer is therefore 63

2. $25 - 5 \div (3 + 2)$

   (a) Start with brackets $(3 + 2) = 5$

   (b) Your sum now reads $25 - 5 \div 5$

   (c) Division comes before subtraction $5 \div 5 = 1$

   (d) Your sum now reads $25 - 1$
       The answer is therefore 24

3. $10 + 6 \times (1 + 10)$

   (a) Start with brackets $(1 + 10) = 11$

   (b) your sum now reads $10 + 6 \times 11$

   (c) Multiplication comes before addition $6 \times 11 = 66$

   (d) Your sum now reads $10 + 66$
       The answer is therefore 76

4. $5(3 + 2) + 5^2$

   (a) Recall that $a(b) = a \times b$
       Hence $5(3 + 2) = 5 \times (3 + 2)$

   (b) That gives you $5 \times 5 + 5^2$

   (c) The next step is orders, in this case, the square $5^2 = 5 \times 5 = 25$. Now you have $5 \times 5 + 25$.

(d) Division and multiplication come before addition and subtraction, so your next step is $5 \times 5 = 25$. Your sum now reads $25 + 25 = 50$.

The answer is 50.

5. $(105 + 206) - 550 \div 5^2 + 10$

    (a) Start with brackets $(105 + 206) = 311$

    (b) Your sum now reads $311 - 550 \div 5^2 + 10$

    (c) Next, orders or powers or exponents and roots.

        In this case we have $5^2 = 25$

    (d) The sum now reads $311 - 550 \div 25 + 10$

    (e) Next is division and multiplication. There is no multiplication in the expression, but the division is $550 \div 25 = 22$.

    (f) Your sum now reads $311 - 22 + 10$

    (g) Although you still have two operations left, addition and subtraction rank equally so you just go from left to right. $311 - 22 = 289$, and $289 + 10 = 299$

        The answer is 299.

6. $7 + 7 \div 7 + 7 \times 7 - 7$

    (a) There are no brackets or orders. So start with division and multiplication.

    (b) $7 \div 7 = 1$ and $7 \times 7 = 49$

    (c) The sum now reads $7 + 1 + 49 - 7$

    (d) Now do the addition and subtraction $7 + 1 + 49 - 7 = 50$

        The answer is 50

Exercises 8

Find the value of the following

1. $18 - 6 \times 2$

2. $6 - 18 \div 3$

3. $16 \div 2 - 3 \times 2$

4. $(5 + 3) + 3 \times 5$

5. $4 \times 6 - (7 - 3)$

6. $8 \times 3 - 17 + 15 \div 5$

7. $28 \div 4 + 2 - 2 \times 4$

8. $6 \times 2 - 2 - 40 \div 4$

9. $5 + (3 + 3) \times 5 \times (2 + 1)$

10. $(5 + 11) \div 4 + 2 + (3 \times 3)$

Example   Simplify the following as far as possible.

1. $3x \times 2 + 5x$

2. $4 \times 5p - 3p$

3. $5 \times 6x - 4x \times 0 - 7x \times 4$

4. $4 \times 8x + 7x \times 3$

5. $3 \times 5x + 4x \div 2$

Solution

1. $3x \times 2 + 5x$

    Parentheses(P): There are no parentheses.

    Exponents(E): There are no exponents

    Multiplication and Division (M and D): There is only multiplication. $3x \times 2 = 6x$

    Addition and Subtraction (A and S): We have only addition. $6x + 5x = 11x$.

    The answer is $11x$

2. $4 \times 5p - 3p$

    P $\to$ There are no parentheses

    E $\to$ There are no exponents

    M and D $\to$ There is only multiplication

    $$4 \times 5p = 20p$$

    A and S $\to$ we now add and subtract

    $$20p - 3p = 17p$$

    The answer is $17p$

3. $5 \times 6x - 4x \times 0 - 7x \times 4 \times (3 + 6)$

    (remember that we work from left to right)

    P $\to$ We have just one pair of parentheses.

    $$(3 + 6) = 9$$

    $\therefore$ the expression becomes
    $$5 \times 6x - 4x \times 0 - 7x \times 4 \times 9$$

    E $\to$ There are no exponents

    M and D $\to$ we have three multiplication operations starting from left.

    (a) $5 \times 6x = 30x$

    (b) $4x \times 0 = 0$

    (c) $7x \times 4 \times 9 = 252x$

    $\therefore$ the expression becomes
    $$30x - 0 - 252x$$

    A and S $\to$ There are two subtraction operations. The result is

    $$30x - 0 - 252x = 30x - 252x = -222x$$

    The answer is $-222x$

Exercises 9

Simplify the following as far as possible.

1. $2x \times 3 + 7x$

2. $-2 \times 2p - 4p$

3. $4 \times 7x - 2x \times 0 - 7x \times 2$

4. $3 \times 8x + 7x \times 2$

5. $5 \times 2x + 4x \div 4$

Expansion of algebraic expressions

- Just as in arithmetic $3(4+5) = (3 \times 4) + (3 \times 5)$, so also in algebra $a(x+y) = ax + ay$.

- Just as in arithmetic $3(4-5) = (3 \times 4) - (3 \times 5)$, so also in algebra $a(x-y) = ax - ay$.

- So it means $a(x-y)$ and $ax - ay$ are the same and equal.

- Expansion involves taking an expression from this form: $a(x-y)$ to this form: $ax - ay$.

- When we expand an algebraic expression, most times it will take more space to write it down.

- When we expand an algebraic expression, we must remove **all** the brackets.

Example   Expand the following algebraic expressions.

1. $8(2c + 3d)$

2. $(4x - y)6$

3. $2(3x - y) + 3(x + 2y)$

Solution

1.

$$8(2c + 3d)$$
$$= 8 \times 2c + 8 \times 3d$$
$$= 16c + 24d$$

2.

$$(4x - y)6$$
$$= 6(4x - y)$$
$$= 6 \times 4x - 6 \times y$$
$$= 24x - 6y$$

3.

$$2(3x - y) + 3(x + 2y)$$
$$= (2 \times 3x - 2 \times y) + (3 \times x + 3 \times 2y)$$
$$= 6x - 2y + 3x + 6y$$
$$= 6x + 3x - 2y + 6y$$
$$= 9x + 4y$$

Exercises 10

Expand the following algebraic expressions

1. $10(3x - 5)$

2. $(x - 2)3$

3. $-5(p + 6)$

4. $(-3 - t)(-10)$

5. $(-4x + y)6$

6. $4(3x - 5) + 3(x - 2)$

7. $2(a + b) - 16(-a + b)$

8. $6(4x + z) - 3(3x - 5z)$

9. $9r - 4(3 + r)$

10. $-11x + 3(3x + 3y)$

Complex Algebraic Expansion

- In arithmetic $(2 + 8)(5 + 10) = 2(5 + 10) + 8(5 + 10)$
  So also in algebra,
$$(a + b)(x + y) = a(x + y) + b(x + y)$$

- In arithmetic $(2 - 8)(5 + 10) = 2(5 + 10) - 8(5 + 10)$
  So also in algebra,
$$(a - b)(x + y) = a(x + y) - b(x + y)$$

Example   Expand the following algebraic expressions

1. $(p + q)(r + s)$

2. $(6x - z)(x - 4z)$

Solution

1.

$$(p + q)(r + s)$$
$$= p(r + s) + q(r + s)$$
$$= pr + ps + qr + qs$$

2.

$$(6x - z)(x - 4z)$$
$$= 6x(x - 4z) - z(x - 4z)$$
$$= 6x \times x - 6x \times 4z - z \times x - z \times (-4z)$$
$$= 6x^2 - 24xz - xz + 4z^2$$

Exercises 11

Expand the following algebraic expressions

1. $(2p - q)(r + s)$

2. $(4 + 5a)(3b - a)$

3. $(2x - 7)(y - 1)$

4. $(3w + x)(5y - v)$

5. $(6x - 2y)(2x - 4y)$

6. $(2x + 9)(2y + 4)$

7. $(a - b)(c + d)$

8. $(3 - p)(4p + 3q)$

9. $(5 + 5a)(3b + 3a)$

10. $(-a + b)(a + b)$

Problems whose answers can be further simplified

- Sometimes it is possible to simplify terms in the final answer. See the example below.

Example   Expand $(9 - x)(7 - 3x)$

Solution

$$(9 - x)(7 - 3x)$$
$$= 9(7 - 3x) - x(7 - 3x)$$
$$= 9 \times 7 - 9 \times 3x - x \times 7 - x \times (-3x)$$
$$= 63 - 27x - 7x + 3x^2$$

- the terms which can be simplified are $-27x$ and $-7x$

$$-27x - 7x = -34x$$

$$\therefore = 63 - 34x + 3x^2 \longleftarrow \text{Final Answer}$$

Exercises 12

Expand and simplify the following

1. $(a + 2x)(3a - x)$

2. $(6 - c)(2 + c)$

3. $(10 + a)(3 - a)$

4. $(8 - ab)(2 + ab)$

5. $(2 + x)(3 - 3x)$

6. $(pq + d)(ax + 3d)$

7. $(mn + 1)(2mn - 6)$

8. $(x - 2c)(3x - y)$

9. $(x + a)(x + a)$

10. $(a + 100b)(a - 100b)$

11. $(10 - p)(10 + p)$

12. $(m^2 - n^2)(m^2 + n^2)$

Special algebraic expansions

Example   Expand the following algebraic expressions and write down your result.

1. $(a + b)(a + b)$

2. $(a - b)(a - b)$

3. $(a + b)(a - b)$

Solution

1.

$$(a + b)(a + b)$$
$$= a(a + b) + b(a + b)$$
$$= a \times a + a \times b + b \times a + b \times b$$
$$= a^2 + ab + ba + b^2$$

but $ba = ab$, so

$$= a^2 + ab + ab + b^2$$

as you can see, the result can be further simplified because $ab + ab = 2ab$

$$= a^2 + 2ab + b^2$$

So we can write

$$(a + b)(a + b) = a^2 + 2ab + b^2$$

2.

$$(a - b)(a - b)$$
$$= a(a - b) - b(a - b)$$
$$= a \times a - a \times b - b \times a - b \times (-b)$$
$$= a^2 - ab - ab + b^2$$

the result can be further simplified because $-ab - ab = -2ab$

$$= a^2 - 2ab + b^2$$

So we can write

$$(a - b)(a - b) = a^2 - 2ab + b^2$$

3.

$$(a + b)(a - b)$$
$$= a(a - b) + b(a - b)$$
$$= a \times a - a \times b + b \times a - b \times b$$
$$= a^2 - ab + ab - b^2$$

the result can be further simplified because $-ab + ab = 0$

So we can write

$$(a + b)(a - b) = a^2 + 0 - b^2$$
$$= a^2 - b^2$$

- If we write down all the results we have had so far, we will have:

  1 $(a + b)(a + b) = a^2 + 2ab + b^2$

  2 $(a - b)(a - b) = a^2 - 2ab + b^2$

  3 $(a + b)(a - b) = a^2 - b^2$

- In arithmetic $2 \times 2 = 2^2$, $a \times a = a^2$. So also, in algebra $(a + b)(a + b) = (a + b)^2$

- So we have the result in a neater form as:

  1 $(a + b)^2 = a^2 + b^2 + 2ab$

  2 $(a - b)^2 = a^2 + b^2 - 2ab$

  3 $(a + b)(a - b) = a^2 - b^2$

- Copy the above results into your notebook and memorize them

Example Expand the following algebraic expressions by directly applying the special results.

1. $(x + 3)^2$

2. $(2h - 10k)^2$

3. $(bt + m)(bt - m)$

4. $(h - 3x)(h + 3x)$

Solution

1.

$$(x + 3)^2$$
$$= x^2 + 3^2 + 2 \times 3 \times x$$
$$= x^2 + 9 + 6x$$

2.

$$(2h - 10k)^2$$
$$= (2h)^2 + (10k)^2 - 2 \times (2h) \times (10k)$$
$$= 4h^2 + 100k^2 - 40hk$$

3.

$$(bt + m)(bt - m)$$
$$= (bt)^2 - (m)^2$$
$$= b^2 t^2 - m^2$$

4.

$$(h - 3x)(h + 3x)$$

note that $(h - 3x)(h + 3x)$ is the same as $(h + 3x)(h - 3x)$

$$= h^2 - (3x)^2$$
$$= h^2 - 9x^2$$

Exercises 13

Expand the following algebraic expressions by directly applying the special results

1. $(3 + a)^2$

2. $(10x - 3y)^2$

3. $(m + 3n)^2$

4. $(m + n)(m - n)$

5. $2(a + b)^2$

6. $-3(x + 1)^2$

7. $5(a + 3)(a - 3)$

8. $-7(h - 3k)(h + 3k)$

9. $(p - 3q)(p + 3q)(-2)$

10. $(f - 2g)(f + 2g)(6)$

Solution of Exercises

Exercises 1

1. rk

2. hb

3. cb

4. jkl

5. mn

6. mnp

7. $abcd$

8. b^3

9. k^6

10. m^3

Exercises 2

1. x^2a

2. $5a^2$

3. b^2a

4. a^2t

5. m^2x

6. $16yz^2$

7. $x^2y^2z^2$

8. p^2z

9. ap^3

10. $22a^3b^3$

Exercises 3

1. $6a^2$

2. p^3q

3. $40k$

4. tm^2n

5. $3gh^2$

6. $55st$

7. $a^2b^2c^2$

8. $10ab^2c$

9. $18000ab^3$

10. $55x^2y^2$

Exercises 4

1. a^9

2. b^{10}

3. a^8

4. c^6

5. z^3

6. a^{12}

7. m^4n^3

8. $5a^4$

9. $2b^5$

10. $30y^5$

Exercises 5

1. $4a$

2. $5a$

3. $3x$

4. $\frac{16}{5}$

5. $65b$

6. $1000g$

7. $3x^2y^2$

8. $5xyz$

9. $5xt$

10. $\frac{12}{5}de$

Exercises 6

1. 6

2. $\frac{3y}{x}$

3. $4x$

4. $\frac{7a}{b}$

5. $17m$

6. d

7. $3x$

8. $7x^2$

9. $4x$

10. $\frac{33mn}{8a}$

11. $\frac{z^2}{2}$

12. $5c$

13. $5x$

14. $26b^3$

15. $6a$

16. $40q$

17. $4z$

18. $9ab$

19. abc

20. $5dq$

Exercises 7

1. $8a^3b^4$

2. $\frac{2a}{bc^4}$

3. $\frac{z^{15}}{x^{15}}$

4. $\frac{y^8}{2}$

5. $\frac{b}{ax^3}$

6. $\frac{6a^{18}}{b}$

7. $\frac{16}{h^2k}$

8. $2x^4$

9. 8

10. $\frac{7}{8a^{99}}$

Exercises 8

1. 6

2. 0

3. 2

4. 23

5. 20

6. 10

7. 1

8. 0

9. 95

10. 15

Exercises 9

1. $13x$

2. $-8p$

3. $14x$

4. $38x$

5. $11x$

Exercises 10

1. $300x - 50$

2. $3x - 6$

3. $-5p - 30$

4. $30 + 10t$

5. $6y - 24x$

6. $15x - 26$

7. $18a - 14b$

8. $15x + 21z$

9. $5r - 12$

10. $9y - 2x$

Exercises 11

1. $2pr + 2ps - qr - qs$

2. $12b - 4a + 15ab - 5a^2$

3. $2xy - 2x - 7y + 7$

4. $15wy - 3wv + 5xy - vx$

5. $12x^2 - 24xy - 4xy + 8y^2$

6. $4xy + 8x + 18y + 36$

7. $ac + ad - bc - bd$

8. $12p + 9q - 4p^2 - 3pq$

9. $15b + 15a + 15ab + 15a^2$

10. $b^2 - a^2$

Exercises 12

1. $3a^2 - ax + 6ax - 2x^2$

2. $12 + 4c - c^2$

3. $30 - 7a - a^2$

4. $16 + 8ab - 2ab - a^2b^2$

5. $6 - 3x - 3x^2$

6. $apqx + 3dpq + dax + 3d^2$

7. $2m^2n^2 - 4mn - 6$

8. $3x^2 - xy - 6cx + 2cy$

9. $x^2 + a^2 + 2ax$

10. $a^2 - 10000b^2$

11. $100 - p^2$

12. $m^4 - n^4$

Exercises 13

1. $9 + a^2 + 6a$

2. $100x^2 - 9y^2 - 60xy$

3. $m^2 + 9n^2 + 6mn$

4. $m^2 - n^2$

5. $2a^2 + 2b^2 + 4ab$

6. $-3x^2 - 3 - 6x$

7. $5a^2 - 45$

8. $63k^2 - 7h^2$

9. $18q^2 - 2p^2$

10. $6f^2 - 24g^2$

Chapter 3

Linear equations

Algebraic equations

- The expression $3 + \boxed{} = 5$ is an algebraic sentence. It can also be written as $3 + x = 5$.

- The algebraic sentence is read as: "three plus an unknown number is equal to 5"

- An equation is simply an algebraic sentence that has an **equals sign**.

- Examples of algebraic equations are

1. $3 + a = 20$
2. $a + b = 2$
3. $6 = a + b$
4. $b = 2$

5. $1 = 16 - 3n$

6. $a \times b = 6$

7. $a \div b = 7$

True and false algebraic sentence

- An algebraic sentence such as $3 + x = 18$ may be true or false depending on the value of the unknown.

Example Say whether the following are true or false.

(a) $x + 2 = 5$ when $x = 3$

(b) $3a + 1 = 10$ when $a = 5$

Solution

(a)

$$x + 2 = 5$$

put $x = 3$

$$\text{i.e., } 3 + 2 = 5$$
$$5 = 5$$

this means the algebraic sentence is true.

(b)

$$3a + 1 = 10$$
$$\text{put } a = 5$$
$$3(5) + 1 = 10$$
$$15 + 1 = 10$$
$$16 \neq 10$$

This means the algebraic sentence is false

(Note $\neq$ means 'not equal to')

Exercises 1

Say whether the following are true or false

1. $p + 3 = 7$ when $p = 4$

2. $5x = 15$ when $x = 3$

3. $k - 2 = 9$ when $k = 10$

4. $11 + d = 16$ when $d = 7$

5. $\frac{12}{x} = 4$ when $x = 3$

6. $25 = 2\frac{1}{2}x$ when $x = 10$

7. $6 = \frac{x}{3}$ when $x = 12$

8. $10 \times l = 10$ when $l = 1$

9. $12 = 12 - m$ when $m = 12$

10. $-19 = 19 + x$ when $x = 0$

Solution of an Equation

- The value of the unknown in an equation which makes the equation true is called the solution of the equation.

- For example, consider this sentence
$$x + 2 = 5$$

- If we put $x = 1$, we get
$$1 + 2 = 3, \text{ but } 3 \neq 5$$

- Since $3 \neq 5$, then $x = 1$ is not the solution

- If we put $x = 3$, we get
$$3 + 2 = 5 \text{ and } 5 = 5$$

- Since $5 = 5$, then $x = 3$ is the solution

Solving Equations by Knowledge of Numbers

Example   Solve the equation $x + 1 = 6$

Solution

- The solution is a number which when 1 is added to equals 6

- Think of such a number

- The answer is 5, because $5 + 1 = 6$

- The solution of the equation $x + 1 = 6$ is therefore 5

$$\text{i.e. } x = 5$$

Example   Solve the equation $18 - x = 7$

Solution

- What number should be subtracted from 18 to get 7

- Think of such a number

- The answer is 11, because $18 - 11 = 7$

- Therefore the solution of $18 - x = 7$ is 11

$$\text{i.e. } x = 11$$

Exercises 2

Solve the following equations using your knowledge of numbers

1. $a + 6 = 7$

2. $5 + x = 10$

3. $8 - y = 4$

4. $10 - x = 9$

5. $z - 20 = 20$

Solving Equations by the Balance Method

- The Balance method is one of the methods of solving an algebraic equation

- There is only one rule applied in the Balance method:
  "whatever operation performed on the LHS must be performed on the RHS"

- 'LHS' is an abbreviation for 'Left-hand Side of Equation'

- 'RHS' is an abbreviation for 'Right-hand Side of Equation'

- The LHS is everything in an equation that comes before the equals sign

- The RHS is everything in an equation that comes after the equals sign

Example

1. $\underbrace{1+1}_{\text{LHS}} = \underbrace{2}_{\text{RHS}}$

2. $\underbrace{a+b}_{\text{LHS}} = \underbrace{c}_{\text{RHS}}$

3. $\underbrace{2}_{\text{LHS}} = \underbrace{x+3-5}_{\text{RHS}}$

4. $\underbrace{2+5+x}_{\text{LHS}} = \underbrace{72+p}_{\text{RHS}}$

- 'LHS' simply refers to everything on the left of the 'equal to' sign

- 'RHS' simply refers to everything on the right of the equal to sign

Demonstration of the Balance Law

- In this section, we shall demonstrate the balance law using some simple expressions

Example

$$1 + 1 = 2$$

- We are going to demonstrate how the balance law works using the arithmetic expression above.

- According to the Balance Law so long as we do the same operation on both sides, the equation remains true.

- This means any operation performed on the LHS must be performed on the RHS.

- In this expression $1 + 1 = 2$, add 10 to both sides.

$$1 + 1 + 10 = 2 + 10$$
$$2 + 10 = 12$$
$$12 = 12$$

- Since $12 = 12$, it means the Balance Law is correct.

- We added 10 to both sides of the equation and the equation remained true because $12 = 12$

- Now subtract 1 from both sides

$$1 + 1 - 1 = 2 - 1$$
$$1 = 1$$

- Again, the equation remains true after performing the same operation on both sides

Example

- $8 = 10 - 2$

- according to the balance law so long as we perform the same operation on both sides the equation remains true

- multiply both sides by 2

$$8 \times 2 = (10 - 2) \times 2$$
$$16 = 8 \times 2$$
$$16 = 16$$

- Since $16 = 16$, the equation remained true after applying the Balance Law

- divide both sides by 4

$$\frac{8}{4} = \frac{(10 - 2)}{4}$$
$$2 = \frac{8}{4}$$
$$2 = 2$$

- Since $2 = 2$, the equation remained true after applying the Balance Law

Example

- In the following examples we shall apply the Balance Law to algebraic equations in order to find their solutions

- Solve the following equations by applying the Balance Law

1. $3 + x = 4$
2. $x - 2 = 5$
3. $8 + y = 16$
4. $b + 16 = 20$
5. $z - 200 = 20$
6. $m + 33 = 3$

Solution

1. $3 + x = 4$

    - We must find a way to make x the only term in the LHS
    - Think. What operation should we perform on both sides to make x to be alone on the LHS?
    - The right operation is this: subtract 3 from both sides

$$3 + x - 3 = 4 - 3$$

- rearrange and evaluate

$$3 - 3 + x = 1$$
$$0 + x = 1$$
$$x = 1$$

- the solution is therefore, $x = 1$

2. $x - 2 = 5$

   - We must find a way to make x the only term in the LHS
   - What operation should we perform on both sides to make x alone?
   - Add 2 to both sides

$$x - 2 + 2 = 5 + 2$$
$$x + 0 = 7$$
$$x = 7$$

   - the solution is therefore, $x = 7$

3. $8 + y = 16$

   - We must find a way to make y the only term in the LHS
   - What operation should we perform on both sides to make y alone?
   - Subtract 8 from both sides

$$8 + y - 8 = 16 - 8$$

   - rearrange and evaluate

$$8 - 8 + y = 16 - 8$$
$$0 + y = 8$$
$$y = 8$$

   - the solution is therefore, $y = 8$

4. $b + 16 = 20$

   - We must find a way to make b the only term on the LHS
   - subtract 16 from both sides

$$b + 16 - 16 = 20 - 16$$
$$b + 0 = 4$$
$$b = 4$$

   - the solution is $b = 4$

5. $z - 200 = 20$

- We must find a way to make z the only term on the LHS

- add 200 to both sides

$$z - 200 + 200 = 20 + 200$$
$$z + 0 = 220$$
$$z = 220$$

- the solution is $z = 220$

6. $m + 33 = 3$

- We must find a way to make m the only term on the LHS

- subtract 33 from both sides

$$m + 33 - 33 = 3 - 33$$
$$m + 0 = -30$$
$$m = -30$$

- the solution is $m = -30$

- Notice that the solution here is negative. The solution of a problem may be positive or negative depending on the problem.

Exercises 3

Solve the following equations.

1. $5 + z = 10$

2. $x - 2 = 20$

3. $b - 20 = 40$

4. $100 + y = 2830$

5. $m - 6 = 12$

6. $a + 20 = 15$

7. $x + 60 = 20$

8. $m - 6 = -5$

9. $8 - n = 16$

10. $z - 11 = -22$

Example Solve the following equations by applying the balance law.

1. $5x = 10$

2. $2a = 16$

3. $3y = 9$

4. $\frac{x}{6} = 3$

5. $\frac{a}{10} = 9$

6. $\frac{z}{6} = 100$

7. $-6z = 36$

8. $11k = -242$

9. $\frac{b}{-8} = 6$

10. $\frac{m}{4} = -60$

Solution

1. $5x = 10$

   - We must make x the only term on the LHS
   - divide both sides by 5

$$\frac{5x}{5} = \frac{10}{5}$$
$$x = 2$$

   - the solution of the equation is $x = 2$

2. $2a = 16$

   - We must make a the only term on the LHS
   - divide both sides by 2

$$\frac{2a}{2} = \frac{16}{2}$$
$$a = 8$$

   - the solution of the equation is $a = 8$

3. $3y = 9$

   - We must make y the only term on the LHS
   - divide both sides by 3

$$\frac{3y}{3} = \frac{9}{3}$$
$$y = 3$$

   - the solution is $y = 3$

4. $\frac{x}{6} = 3$

   - We must make x the only term on the LHS
   - Multiply both sides by 6

$$\frac{x}{6} \times 6 = 3 \times 6$$
$$x = 18$$

   - the solution is $x = 18$

5. $\frac{a}{10} = 9$

   - We must make a the only term on the LHS

- Multiply both sides by 10

$$\frac{a}{10} \times 10 = 9 \times 10$$
$$a = 90$$

- the solution is $a = 90$

6. $\frac{z}{6} = 100$

    - We must make z the only term on the LHS
    - Multiply both sides by 6

$$\frac{z}{6} \times 6 = 100 \times 6$$
$$z = 600$$

- the solution is $z = 600$

7. $-6z = 36$

    - We must make z the only term on the LHS
    - divide both sides by -6

$$\frac{-6z}{-6} = \frac{36}{-6}$$
$$z = -6$$

- the solution is $z = -6$

8. $11k = -242$

    - We must make k the only term on the LHS
    - divide both sides by 11

$$\frac{11k}{11} = \frac{-242}{11}$$
$$k = -22$$

- the solution is $k = -22$

9. $\frac{b}{-8} = 6$

    - We must make b the only term on the LHS
    - Multiply both sides by -8

$$\frac{b}{-8} \times -8 = 6 \times (-8)$$
$$b = -48$$

- the solution is $b = -48$

10. $\frac{m}{4} = -60$

- We must make m the only term on the LHS
- Multiply both sides by 4

$$\frac{m}{4} \times 4 = -60 \times 4$$
$$m = -240$$

- the solution is $m = -240$

Exercises 4

Solve the following equations

1. $4a = 160$
2. $13x = 132$
3. $\frac{m}{6} = 13$
4. $\frac{k}{8} = 10$
5. $x \div 18 = 36$

6. $7 \times a = 84$
7. $y \times 11 = 99$
8. $k \times -6 = 180$
9. $b \div (-2) = 16$
10. $\frac{x}{5} = -205$

Example   Solve the following equations

1. $2 - a = 4$
2. $-10 - x = 22$
3. $-p + 8 = 3$
4. $-k - 2 = -6$

Solution

1. $2 - a = 4$

- We must make a the only term on the LHS
- Subtract 2 from both sides

$$2 - a - 2 = 4 - 2$$

- rearrange and evaluate

$$2 - 2 - a = 4 - 2$$
$$0 - a = 2$$
$$-a = 2$$

- We have not arrived at the solution yet because we have $-a$ on the LHS instead of a
- In order to get 'a' on the LHS, we must divide both sides by -1

$$\frac{-a}{-1} = \frac{2}{-1}$$
$$a = -2$$

- the solution is $a = -2$

2. $-10 - x = 22$

- We must make x the only term on the LHS
- add 10 to both sides

$$-10 - x + 10 = 22 + 10$$

- rearrange and evaluate

$$-10 + 10 - x = 22 + 10$$
$$0 - x = 32$$
$$-x = 32$$

- In order to get x on the LHS we must divide both sides by -1

$$\frac{-x}{-1} = \frac{32}{-1}$$
$$x = -32$$

- the solution is $x = -32$

3. $-p + 8 = 3$

- We must make p the only term on the LHS
- Subtract 8 from both sides

$$-p + 8 - 8 = 3 - 8$$
$$-p + 0 = -5$$
$$-p = -5$$

- In order to make p the only term on the LHS we must divide both sides by -1

$$\frac{-p}{-1} = \frac{-5}{-1}$$
$$p = 5$$

- the solution is $p = 5$

4. $-k - 2 = -6$

- We must make k the only term on the LHS

- add 2 to both sides

$$-k - 2 + 2 = -6 + 2$$
$$-k + 0 = -4$$
$$-k = -4$$

- In order to get k as the only term on the LHS we must divide both sides by -1

$$\frac{-k}{-1} = \frac{-4}{-1}$$
$$k = 4$$

- the solution is $k = 4$

Exercises 5

Solve the following equations

1. $-z + 200 = -20$
2. $-m - 30 = 60$
3. $6 - a = 12$
4. $-b - 16 = -20$
5. $20 - x = 0$

6. $3 = x + 8$
7. $5 = g - 11$
8. $-x - 3 = 1$
9. $20 = -a - 6$
10. $15 = 16 - m$

Example   Solve the following equations

1. $2x + 4 = 20$
2. $8y + 1 = 17$
3. $-6z + 39 = 3$

Solution

1. $2x + 4 = 20$

    - We must make x the only term on the LHS
    - subtract 4 from both sides

    $$2x + 4 - 4 = 20 - 4$$
    $$2x + 0 = 16$$
    $$2x = 16$$

    - divide both sides by 2

    $$\frac{2x}{2} = \frac{16}{2}$$
    $$x = 8$$

- the solution is $x = 8$

2. $8y + 1 = 17$

  - We must make y the only term on the LHS
  - subtract 1 from both sides

$$8y + 1 - 1 = 17 - 1$$
$$8y = 16$$

  - divide both sides by 8

$$\frac{8y}{8} = \frac{16}{8}$$
$$y = 2$$

  - the answer is $y = 2$

3. $-6z + 39 = 3$

  - We must make z the only term on the LHS
  - subtract 39 from both sides

$$-6z + 39 - 39 = 3 - 39$$
$$-6z + 0 = -36$$
$$-6z = -36$$

  - divide both sides by -6

$$\frac{-6z}{-6} = \frac{-36}{-6}$$
$$z = 6$$

  - the solution is $z = 6$

Exercises 6

Solve the following equations

1. $2a + 6 = 10$

2. $11m + 3 = 124$

3. $-5 + 7y = 9$

4. $6p + 3 = 15$

5. $3m - 4 = 26$

6. $5p - 30 = 0$

7. $4 = 4 + 8w$

8. $9 = 5x - 1$

9. $4 + 3d = 25$

10. $8x - 9 = 7$

Problems with Fractional Answers

Example Solve the following equations

1. $8x = 17$

2. $3y + 8 = 42$

Solution

1.

$$8x = 17$$

- divide both sides by 8

$$\frac{8x}{8} = \frac{17}{8}$$
$$x = \frac{17}{8}$$

- the answer is $\frac{17}{8}$

2.

$$3y + 8 = 42$$

- subtract 8 from both sides

$$3y + 8 - 8 = 42 - 8$$
$$3y + 0 = 34$$
$$3y = 34$$

- divide both sides by 3

$$\frac{3y}{3} = \frac{34}{3}$$
$$y = \frac{34}{3}$$

- the answer is $\frac{34}{3}$

Exercises 7

Solve the following equations

1. $7p - 9 = 18$

2. $21 = 8y - 4$

3. $4a + 3 = 16$

4. $10m - 11 = 12$

5. $6s - 5 = 16$

6. $21 = 4 + 9n$

Checking the Solution

- It is possible to check if a solution is correct.

- For example, if we solve

$$2x + 1 = 5$$

- subtract 1 from both sides

$$2x + 1 - 1 = 5 - 1$$
$$2x + 0 = 4$$
$$2x = 4$$

- divide both sides by 2

$$\frac{2x}{2} = \frac{4}{2}$$
$$x = 2$$

- the answer is $x = 2$

- Now to check if this solution is correct simply put $x = 2$ in the LHS of the initial question.

- That is,

$$2(2) + 1$$
$$= 2 \times 2 + 1$$
$$= 4 + 1$$
$$= 5$$
$$\therefore \text{ LHS} = \text{RHS}$$

- hence, we have proven that the answer $x = 2$ is correct, because LHS = RHS when $x = 2$.

Example   Solve the following equations. After solving them check if your answer is correct.

1. $5y + 6 = 1$

2. $4 = 4 + 8w$

Solution

1.

$$5y + 6 = 1$$

- subtract 6 from both sides

$$5y + 6 - 6 = 1 - 6$$
$$5y + 0 = -5$$
$$5y = -5$$

- divide both sides by 5

$$\frac{5y}{5} = \frac{-5}{5}$$
$$y = -1$$

- the answer is $y = -1$
- **Check answer:** Check your answer by putting $y = -1$ in the initial question
- that is,

$$5(-1) + 6$$
$$= -5 + 6$$
$$= 1$$
$$\therefore \text{ LHS} = \text{RHS}$$

- thus the solution $y = -1$ is correct

2.

$$4 = 4 + 8w$$

- subtract 4 from both sides

$$4 - 4 = 4 + 8w - 4$$
$$0 = 8w$$

- divide both sides by 8

$$\frac{0}{8} = \frac{8w}{8}$$
$$0 = w$$
$$w = 0$$

- the answer is $w = 0$
- Check your answer by putting $w = 0$ in the initial question
- that means

$$4 = 4 + 8(0)$$
$$4 = 4 + 0$$
$$4 = 4$$

- therefore the solution is correct

Exercises 8

Solve the following equations and check your solution if it is correct

1. $3x + 4 = 17$

2. $5x + 8 = 11$

3. $3 + 2y = 10$

4. $19 = 6 + 9x$

5. $10x - 3 = 5$

6. $6t - 5 = 6$

7. $4s - 1 = 2$

8. $2 = 7m - 4$

9. $8n + 3 = 4$

10. $12 + 3a = 21$

Solution of Exercises

Exercises 1

1. true

2. true

3. false

4. false

5. true

6. true

7. false

8. true

9. false

10. false

Exercises 2

1. $a = 1$

2. $x = 5$

3. $y = 4$

4. $x = 1$

5. $z = 40$

Exercises 3

1. $z = 10$

2. $x = 22$

3. $b = 60$

4. $y = 2730$

5. $m = 18$

6. $a = -5$

7. $x = -40$

8. $m = 1$

9. $m = -8$

10. $n = -16$

11. $z = -11$

Exercises 4

1. $a = 40$

2. $x = \frac{132}{13}$

3. $m = 78$

4. $k = 80$

5. $x = 648$

6. $a = 12$

7. $y = 9$

8. $k = -30$

9. $b = 32$

10. $x = -1025$

Exercises 5

1. $z = 220$

2. $m = -90$

3. $a = -6$

4. $b = 4$

5. $x = 20$

6. $x = -5$

7. $g = 16$

8. $x = -4$

9. $a = -26$

10. $m = 1$

Exercises 6

1. $a = 2$

2. $m = 11$

3. $y = 2$

4. $p = 2$

5. $m = 10$

6. $p = 6$

7. $w = 0$

8. $x = 2$

9. $d = 7$

10. $x = 2$

Exercises 7

1. $p = \frac{27}{7}$

2. $y = \frac{25}{8}$

3. $a = \frac{13}{4}$

4. $m = \frac{23}{10}$

5. $s = \frac{21}{6}$

6. $n = \frac{17}{9}$

Exercises 8

1. $x = \frac{13}{3}$

2. $x = \frac{3}{5}$

3. $y = \frac{7}{2}$

4. $x = \frac{13}{9}$

5. $x = \frac{8}{10}$

6. $t = \frac{11}{6}$

7. $s = \frac{3}{4}$

8. $m = \frac{6}{7}$

9. $n = \frac{1}{8}$

10. $a = 3$

Chapter 4

Word problems

Arithmetic word problems for addition and subtraction

Example A boy has five balls. If he lends two to his friend, how many does he have left?

Solution

- By simple reasoning you know that the boy will have 3 balls left

- But why does the boy have 3 balls left?

- Can you prove with a simple equation that the boy has 3 balls left?

- The simple equation which proves that the boy has three balls left is

$$5 - 2 = 3$$

- Can you create similar equation for other word problems

- Now study the following example

Example Richard has two pencils. He bought five more pencils from the local shop. How many pencils does Richard have now?

Solution

- By simple reasoning you know that Richard has 7 pencils now

- Can you write an equation to prove this?

- The simple equation which proves this is

$$2 + 5 = 7$$

- This is because Richard initially had 2 pencils

- Then he bought 5 more, i.e. +5 pencils

- So altogether Richard has $2 + 5$ pencils i.e. 7 pencils

Exercises 1

Answer the following questions by simple reasoning and write the equation to prove your answer.

1. A boy has five milk bottles. If he sells two, how many does he have left?

2. Baba is 10 years old. How old will he be in 2 years time.

3. A farmer has 4 kola nuts. A friend gives him two more kola nuts. How many kola nut does he have now?

4. A man has eight T-shirts. How many more T-shirts must he buy to have ten T-shirts?

5. Five eggs were removed from a basket containing 10 eggs. How many eggs are left in the basket?

Algebra word problems for addition and subtraction

- Sometimes a word problem cannot be solved by simple reasoning with knowledge of numbers.

- In this section we introduce a better way of solving word problems by employing algebra.

- The unknown in any word problem will be represented by z. (You can use any letter, but we choose z for convenience)

- Then we write the equation for the problem and solve for z.

Example   John sells oranges in the local shop. On Monday he sells 15 oranges. On Tuesday he sells 11 oranges. On Wednesday he sells just 2 oranges. How many oranges has he sold altogether?

Solution

(i) First we must identify the unknown quantity. The unknown quantity is

'Number of oranges sold altogether'

(ii) Now we shall assign a letter to represent the unknown quantity. Let the letter be z. (You can assign any letter, but we will be using z to represent the unknown quantity henceforth)

This means z = number of oranges sold altogether

(iii) Write the equation
$$z = 15 + 11 + 2$$

(iv) Evaluate z

$$z = 26 + 2 = 28$$
$$z = 28 \quad \text{oranges}$$

$\therefore$ number of oranges sold altogether is 28

Example A book has 100 pages. Samuel reads 16 pages in the morning, 10 pages in the afternoon and the remaining pages in the evening.
How many pages did he read in the evening?

Solution

(i) The unknown quantity is 'number of pages read in the evening'

(ii) Let z = number of pages read in the evening

(iii) Write the equation

$$z = 100 - 16 - 10$$

(iv) Evaluate z

$$z = 84 - 10$$
$$z = 74$$

∴ the number of pages Samuel read in the evening is 74

Exercises 2

1. A farm contains 54 pigs. The farmer bought another 70 pigs to add to the farm livestock. How many pigs are there in the farm?

2. The largest toy soldier battle in the city involved James' 30 soldiers, Peter's 39 soldiers, Kalu's 88 soldiers and Lucky's 53 soldiers. How many soldiers participated in the battle?

3. John is 12 years old. Mike is 18 years old. If the age of Tom is equal to the difference between Mike and John's age, how old is Tom?

4. A man loads three bags of different colours with coins. The red bag contained 100 coins, the blue bag contained 16 coins and the yellow bag contained 115 coins. If he loses the blue bag how many coins does he have?

5. A tree has 50 fruits. If 6 fell in the morning, 13 fell in the afternoon and 30 fell in the evening. How many fruits will be left in the tree at the end of the day?

6. In a poultry, there are 300 birds. On Monday 15 died and 11 were born. On Tuesday 3 died and 20 were born. How many birds are alive now?

Addition and subtraction key words used in algebra

In order to know whether a word problem involves addition or subtraction of quantities involved you may memorize the following tip. This will help you to easily translate word problems into equations

Addition key words

1. add

2. all together (or altogether)

3. plus

4. sum

5. and

6. combined

7. both

8. in all

9. increased by

10. more than

11. together

12. total

13. how many in all

14. how much

Subtraction key words

1. subtract

2. difference

3. minus

4. take away

5. reduce

6. less or less than

7. fewer or fewer than

8. lost

9. change in quantity (e.g. change in price, change in height e.t.c.)

10. how much is left

The following problems will use some key words listed above

Exercises 3

1. What number do you get when you **add** 1083 to 2102

2. There are 5 oranges in a basket. Clinton puts 13 apples in the same basket. How many fruits are there **all together** in the basket?

3. A phone is worth $12 and a TV set is worth $200. How much is **both** worth?

4. What is the **sum** of 2299 and 1083?

5. Rita has 5 pairs of shoes. If Martha has six pairs of shoes **more than** Rita, how many shoes does Martha have?

Exercises 4

1. What is the result if 3085 is **subtracted** from 5000

2. The price of a watch is $60. The price of a necklace is $100. What is the **difference** in price between a watch and a necklace.

3. Tom has to walk a total distance of 20 km to get to school. If he walks 6 km, **how much longer** does he have to walk before he gets to school?

4. A forest contains $2,566$ trees. If only 322 trees survived the drought, how many trees were **lost**.

5. Evaluate the money in Jack's account **minus** the money he spent. He spent $60 and initially had $1000

Algebra Word Problems (Multi-step Problems)

Example In a test James got 7 marks more than Emily. Emily got 10 marks. How many marks did James get?

Solution

(i) The unknown is "James' mark"

(ii) Let $z =$ James' mark

(iii) From the first part of the question, James got 7 marks more than Emily. If we represent Emily's mark by e then
$$z = e + 7$$

(iv) In the second part of the question it is revealed that Emily got 10 marks, hence $e = 10$.

(v) Put $e = 10$ in the equation:

$$z = 10 + 7$$
$$z = 17$$

$\therefore$ James got 17 marks

Example Jude has $12 less than Andrew. Andrew has $20. How much does Jude have?

Solution

(i) The unknown is "How much Jude has"

(ii) Let $z =$ how much Jude has

(iii) From the first part of the question, Jude has $12 less than Andrew. Let us represent Andrew's money with a.
$$\text{then} \quad z = a - 12$$

(iv) In the second part of the question, it is stated that Andrew has $20. Hence $a = 20$

(v)

$$z = 20 - 12$$
$$z = \$8$$

$\therefore$ James has $8

Example There are 10 balls in a basket. Peter took some balls from the basket, leaving 3 balls. How many balls did Peter take?

Solution

(i) The unknown quantity is 'the number of balls Peter took'

(ii) Let $z =$ the number of balls Peter took

(iii) In the first part of the question it is stated that the total number of balls is 10.

(iv) The second part of the problem states that the remaining balls are 3.

$$\text{But} \quad z + 3 = \text{total number of balls}$$
$$\text{hence} \quad z + 3 = 10$$

Subtract 3 from both sides

$$z = 7$$

Example Rachel sells apples. On Monday she sold 10 apples. On Tuesday she sold 16 more apples than she did on Monday.

1. How many apples did she sell on Tuesday?

2. How many apples did she sell altogether?

Solution

1. (i) The unknown is the number of apples she sold on Tuesday

   (ii) Let $z =$ number of apples sold on Tuesday

   (iii)

$$z = 16 + 10$$
$$z = 26$$

2. (i) The unknown is the total number of apples sold.

   (ii) Let $z =$ total number of apples sold

   (iii)

$$z = (\text{apples sold on Monday}) + (\text{apples sold on Tuesday})$$
$$z = 10 + 26$$
$$z = 36$$

$\therefore$ 36 apples were sold in total.

Exercises 5

1. John's school bag contains three different kinds of items: notebooks, pencils and erasers. Before John went to schocl he had 10 notebooks, 5 pencils and 6 erasers. During school hours he gave some items to his friend Paul. He came back home from school with a total of 10 items. How many items did John give Paul?

2. After giving $500 to his son. Mr Jack had $2100 left in his wallet. How much money did he have at first?

3. Lily has 22 pencils. Sandra has 5 pencils more than Lily. James has 6 pencils less than Sandra. Ben has 10 pencils more than James.

   (a) What is the sum of number of pencils owned by Sandra and James?
   (b) What is the sum of number of pencils owned by Lily and James?
   (c) What is the total number of pencils?

4. The total number of students in a class is 33. If the number of boys is 13, how many girls are present in the class?

5. A box contains 40 cards of five different colours: white, green, yellow, black and red. Find the number of white cards if the number of green, yellow, black and red cards are $5, 10, 15$ and 2 respectively.

Algebra Word Problems (Unknowns are Present in the Question)

In this type of problem, apart from z we have another unknown in the problem. The other unknown is already assigned a letter in the problem. The solution to this type of problem does not give a number result such as $20, 5, 6, 100$ e.t.c. Instead, the result is an expression such as $14 - x$, $2 + a$, $5 + y$, e.t.c.

Example A boy has m mangoes. He sells two of them. How many mangoes does he now have?

Solution

(i) The unknown is 'the number of mangoes the boy has presently'

(ii) Let $z =$ 'number of mangoes the boy has presently'

(iii) Write the equation
$$z = m - 2 \leftarrow \text{Final Answer}$$

Hence the boy now has $m - 2$ mangoes

Example The rainforest near the town contains 10 mango trees, 6 Iroko trees, a pine trees, and b coconut trees. How many trees are there in the rainforest altogether.

Solution

(i) The unknown is 'the number of trees altogether'

(ii) Let $z =$ 'the number of trees altogether'

(iii) Write the equation

$$z = 10 + 6 + a + b$$
$$z = 16 + a + b \leftarrow \text{Final answer}$$

Example Each side of a square board is d centimeters long. What is the perimeter of the board?

Solution

(i) The unknown quantity is 'the perimeter of the board'

(ii) Let $z =$ 'the perimeter of the board'

(iii) Write the equation

$$z = d + d + d + d$$
$$z = 4d$$

Note Perimeter is the sum of the distances around the boundary of a plane shape.

Example A girl is 14 years old today. How old will she be in x years time

Solution

(i) The unknown is the girl's future age

(ii) Let $z =$ The girl's future age

(iii) Write the equation

$$z = 14 + x$$

∴ The answer is $14 + x$ years old

Exercises 6

1. Baba is 10 years old. How old will he be y years from now?

2. A smartphone has 12 user apps installed. If there are y system apps, how many apps are there altogether?

3. In a test, Richard got 20 marks more than John. John got x marks. How many marks did Richard get?

4. A boy has a oranges, b apples and 4 mangoes. How much fruit does he have altogether?

5. A woman is 30 years old. What was her age y years ago?

6. A book has w pages. A boy reads 30 pages of the book. How many pages has the boy not read?

7. A girl guesses that a line is $15\,\text{cm}$ long. She measures the line. She finds that her guess is $x\,\text{cm}$ too long. What is the length of the line?

8. A dealer bought a diamond for $\$x$ and sold it for \$165500. What was the profit?

9. A man cycles $x\,\text{km}$ towards a village which is $20\,\text{km}$ away. How far does he still have to cycle to get to the village?

10. Audu has s dollars in his savings account. He uses some of this to buy a suit which costs \$100. How much does he have left in his savings?

11. A farmer has 6 goats. If he buys m cows, n hens and 5 pigs. How many animals does he have in total?

12. A tree has 12 fruits. If m fruits are harvested in the morning, n fell in the afternoon and 6 fell in the evening, how many fruits will be left on the tree at the end of the day?

13. Peter took 5 pencils from a box containing k pencils. John put 25 more pencils in the basket, but Timmy took h pencils from the basket. How many pencils are left in the basket?

14. Lily has n pencils. Sandra has 5 pencils more than Lily. James has p pencils less than Sandra. How many pencils do they have altogether?

15. A box contains 40 cards of red, yellow and blue colour. If t cards are red and 6 cards are yellow, how many cards are blue?

Multiplication word problems

Key Words Used in Multiplication Word Problem

1. multiplied by

2. by

3. double

4. triple

5. each group

6. every

7. factor of

8. increased by

9. multiplied by

10. of

11. product

12. times

13. twice

14. thrice

Multiplication word problems (One-Step Problems)

Example   A table costs \$50. What is the cost of 10 such tables

Solution

(i) The unknown is 'the cost of 10 tables'

(ii) Let $z =$ 'the cost of 10 tables'

(iii) Write the equation

$$z = 50 \times 10$$
$$z = \$500$$

∴ the cost of 10 such tables is $500

Example Jackson delivers 412 newspapers in a day. How many newspapers does he deliver in 6 days?

Solution

(i) The unknown is 'the number of newspapers he delivers in 6 days'

(ii) Let $z =$ 'the number of newspapers he delivers in 6 days'

(iii) Write the equation

$$z = 412 \times 6$$
$$z = 2472 \text{ newspapers}$$

∴ he delivers 2472 newspapers in 6 days

Example John bought two boxes of books from a local shop. If each box had five books, how many books did he buy?

Solution $z = 2 \times 5 = 10$ books.

Example Everybody in Richard's class each gave him five dollars. If Richard has 20 classmates, how many dollars does he now have?

Solution $z = 5 \times 20 = \$100$

Exercises 7

1. If there are 7 days in a week, how many days are there in five weeks?

2. There are 366 days in a year. How many days are in 15 years?

3. What is the cost of 5 books at $30 each?

4. What number is 9 times greater than 5?

5. An employee at a construction site earns $8 an hour. If he works eight hours in one week. How much would he have earned?

6. Bisi has 10 chocolate boxes. Each chocolate box has 16 chocolate bars in it. How many chocolate bars does Bisi have in total?

Multiplication word problems (Two-step Problems)

Example A coat costs 5 times as much as a pair of trousers. If the pair of trousers costs $86, what is the total cost of the coat and the pair of trousers?

Solution

(i) The unknown is 'the total cost of coat and a pair of trousers'

(ii) Let $z =$ 'the total cost of coat and a pair of trousers'

(iii) From the first part of the problem it is stated that a coat costs 5 times as much as a pair of trousers. But a pair of trousers is $86. Hence

$$\text{'price of coat'} = 5 \times 86$$
$$= \$430$$

(iv)

$$z = (\text{price of coat}) + (\text{price of a pair of trousers})$$
$$z = 430 + 86$$
$$z = \$516$$

Example Betty needs 2245g of sugar to bake some cookies. She has 4 packets of sugar. The mass of each packet is 500g. How much more sugar does she need?

Solution

- 4 packets of sugar is

$$4 \times 500\text{g}$$
$$= 2000\text{g}$$

- She needs 2245g of sugar

- How much more does she need?

$$2245 - 2000 = 245\text{g}$$

Exercises 8

1. Susan has 400 sheets of paper. She gives 5 sheets of paper to every student in a class. There are 62 students in the class. How many sheets of paper has she left?

2. Sally enters an eating competition with her friend Josie. Sally eats 27 hot dogs in two hours. Josie eats seven times more hot dogs than sally does.

   (a) How many hot dogs does Josie eat?
   (b) How many more hot dogs does Josie eat than Sally?

3. Jude went running 4 days this past week. He ran 9km each day. In that same week, Bethel ran 15 fewer km than Jude. How many kilometers did Bethel run that week?

4. What is the difference between 100 and a number which is nine times greater than 6.

5. Bisi has 16 chocolate boxes. Each chocolate box contains 100 chocolate bars. If she sold 5 chocolate boxes, how many chocolate bars does she have now.

Multiplication Word Problems (unknown is Present in the Question)

Example

1. How many days are there in w weeks

   Solution

   (i) The unknown is the number of days in w weeks

   (ii) Let z = number of days in w weeks

   (iii) Write the equation

   $$z = 7 \times w = 7w$$

   ('7' came as a result of '7 days in one week')

2. What is the cost of x books at \$30 each?

   Solution

   (i) The unknown is 'the cost of x books'

   (ii) Let z = 'the cost of x books'

   (iii) Write the equation.

   $$z = 30 \times x$$
   $$z = 30x$$

3. A girl is x years old. Her father is four times as old. How old is her father?

   Solution

   (i) The unknown quantity is 'the father's age'

   (ii) Let z = "The father's age"

   (iii) Write the equation

   $$z = 4 \times x$$
   $$z = 4x$$

Exercises 9

1. A car costs p when new. It was sold for four-fifths of its cost price. How much money was lost on the car?

2. A girl is x years old. Her father is five times as old. What will be the age of the father in y years time.

3. A man has n cattle ranches. If each cattle ranch contains p cattle, how many cattle does he have?

4. There are 366 days in a year. How many days are in t years?

5. A coat costs p times as much as a pair of trousers. If the pair of trousers cost $100, what is the total cost of the coat and the pair of trousers?

Division word problems

Key Words Used in Division

1. as much
2. equal sharing
3. half
4. quarter
5. percent

6. quotient of
7. how many in each
8. ratio of
9. divide into parts
10. cut up

Division word problems (One-step Problems)

Example

1. There are 160 primary school students in a school. The students are to be equally divided into 5 classes. How many students do we have in each class.

   Solution

   (i) The unknown is 'the number of students in each class'
   (ii) Let z = 'the number of students in each class'
   (iii) Write the equation

$$z = \frac{160}{5}$$
$$z = 32$$

   $\therefore$ there are 32 students in each class

2. Melissa made 50 cupcakes. She packed 10 cupcakes into each box. How many boxes of cupcakes did she pack.

Solution

(i) The unknown quantity is 'the number of boxes of cupcakes'

(ii) Let z = 'the number of boxes of cupcakes'

(iii) Write the equation.

$$z = \frac{50}{10}$$
$$z = 5$$

$\therefore$ there are 5 boxes of cupcakes

3. How many weeks are in 14 days if there are seven days in one week.

Solution

(i) The unknown quantity is 'the number of weeks in 14 days'

(ii) Let z = 'The number of weeks in 14 days'

(iii) Write the equation.

$$z = \frac{14}{7}$$
$$z = 2$$

$\therefore$ there are 2 weeks in 14 days

Exercises 10

1. (i) How many weeks are in 84 days.

   (ii) How many weeks are in 21 days.

2. The total money Richard received from his classmates is \$100. If Richard has 20 classmates how much did each of Richard's classmates give him?

3. How many years are in 732 days?

4. (i) If 8 books cost \$3200, what is the unit price?

   (ii) If 5 books cost \$150, what is the unit price?

5. What number is one-third as big as 9.

Division word problems (Multi-step Problems)

Example   Marcus had 700 marbles. He gave away 175 marbles and put the remaining marbles equally into 5 bags. How many marbles were there in each bag?

Solution

$$\textbf{Step 1:} \quad 700 - 175 = 525$$

$$\textbf{Step 2:} \quad \frac{525}{5} = 105$$

$\therefore$ there were 105 marbles in each bag.

Example Rosaline made 364 donuts. She put 4 donuts into each box.

(a) How many boxes of donuts were there?

(b) If she sold each box for \$3, how much money would she receive?

Solution

(a)

$$z = \frac{364}{4}$$
$$z = 91$$

$\therefore$ there were 91 boxes of donuts.

(b)

$$z = 91 \times 3$$
$$z = \$273$$

$\therefore$ she would receive \$273

Example 34 girls and 24 boys go on a school trip. The children are split into two equal groups. How many children are in each group?

Solution

$$\textbf{Step 1:} \quad z_1 = 34 + 24 = 58$$

$$\textbf{Step 2:} \quad z_2 = \frac{58}{2} = 29$$
$$z_2 = 29$$

$\therefore$ there are **29 children** in each group.

Example Josh bought 45 sweets on Monday and 30 sweets on Tuesday. He shared all the sweets between his five friends. How many sweets did they get each?

Solution

$$\textbf{Step 1:} \quad z_1 = 45 + 30 = 75$$
$$\textbf{Step 2:} \quad z_2 = \frac{75}{5}$$
$$z_2 = 15$$

∴ each of Josh's friend got **15 sweets**.

Exercises 11

1. I mix 50ml of lemonade with 30ml of orange juice and divided the mixture between two glasses. How much drink is in each glass?

2. Mrs Smith buys 42cm of red ribbon and 24cm of blue ribbon. She shares the ribbon equally between her 3 daughters. How much ribbon do they each get?

3. A group of 25 adults and 5 children go to a restaurant. Each table can seat 5 people. How many tables do the group need.

4. Mr Bala had 35 chickens. He sold 3 and kept the rest in 4 equal groups. How many chickens were in each group.

5. Mrs Jones wins €45 on the lottery and €30 on a scratch card. She shares the winning equally between her 3 children. How much do they get each.

Division word problems (Unknown is Present in the Question)

Example I have n biscuits in one box and p biscuits in another. Each child needs 2 biscuits. How many children can I feed.

Solution

$$\textbf{Step 1:} \quad z_1 = n + p$$
$$\textbf{Step 2:} \quad z_2 = \frac{n + p}{2}$$

∴ $\frac{n+p}{2}$ children can be fed

Example Mrs Higgins bakes 15 chocolate cakes and x sponge cakes. She puts them in boxes for a cake sale. Each box can hold n cakes. How many boxes will Mrs Higgins need?

Solution

$$\textbf{Step 1:} \quad z_1 = 15 + x$$
$$\textbf{Step 1:} \quad z_2 = \frac{15 + x}{n}$$

∴ Mrs Higgins needs $\frac{15+x}{n}$ boxes.

Exercises 12

1. 120 boys and 130 girls go on a school trip. They travel on p coaches. How many children are there on each coach?

2. I mix together $x$ml of lemonade and 680ml of orange juice. I divide the drink equally between 8 glasses. How much drink is in each glass?

3. Joe has m Lego bricks. He sold n Lego bricks. He split the remaining bricks equally into p boxes. How many bricks were in each box?

4. A theatre can seat 360 people in the stalls and t people in the balcony. Each row can seat m people. How many rows are there.

5. Mrs Jones wins €x on the lottery one week and €1030 on the lottery the following week. She shares the winnings equally between her n children. How much money do they get each.

Division word problems (Problems Involving Mixed Fractions)

Example   Rosaline made 364 donuts. She put 8 donuts into each box.

(a) How many boxes of donuts were there?

(b) How many donuts were left over?

(c) If she sold each box for $3, how much money would she receive.

Solution

(a)

$$z = \text{number of boxes of donuts}$$
$$z = \frac{364}{8}$$
$$z = 45.5 = 45\frac{1}{2}$$

→ Notice that the result is not a whole number but a mixed fraction.

→ Simply take the whole number part as the answer in such cases.

$$\text{i.e}\quad z = 45$$

∴ there are 45 boxes.

(b) From part (a), since the result was not a whole number, it means some donuts were left over.

To get the number of donuts left over, we will subtract 45×8 from 364.

$$\text{Now}\quad 45 \times 8 = 360$$
$$\text{and}\quad 364 - 360 = 4$$

∴ **4 donuts** were left over

(c) $z =$ how much she will receive

there are 45 boxes, each box is $3,

$$\text{hence} \quad z = 45 \times 3$$
$$z = \$135$$

$\therefore$ she will receive $135.

Example Marcus had 700 marbles. He gave away 175 marbles and put the remaining marbles equally into 4 bags.

(a) How many marbles were there in each bag.

(b) How many marbles were left over

Answer

(a)

$$\textbf{Step 1:} \quad z_1 = 700 - 175 = 525$$
$$\textbf{Step 2:} \quad z_2 = \frac{525}{4} = 131.25$$
$$= 131\frac{1}{4}$$

the mixed fraction is $131\frac{1}{4}$ i.e. $131 + \frac{1}{4}$

the whole number part is 131

the fraction part is $\frac{1}{4}$

$\rightarrow$ the answer is the whole number part

$\therefore$ there were 131 marbles in each bag.

(b) To get the number of leftover marbles we must subtract 131×4 from 525

$$\text{now} \quad 131 \times 4 = 524$$
$$\text{and} \quad 525 - 524 = 1$$

$\therefore$ **1 marble** was left over.

Exercises 13

1. If 18 cookies are arranged equally in 5 piles

    (a) How many cookies are in each pile

    (b) How many cookies are left-over.

2. $50 was shared equally among 3 people. The remaining money was given to Ben. How much did Ben get?

3. Mrs Peters wins $4500 on the lottery and $3100 on a scratch card. She shares her winnings between her 3 children. How much do they get each?

4. Micheal had 35 chickens. He sold 4 and shared the rest to four of his friends, and kept the remainder. How many chickens did Michael keep?

5. Joshua bought 46 sweets on Saturday and 31 sweets on Sunday. He shared all the sweets between his five friends and sold the remainder for $10 each. How much did he make?

Solution of Exercises

Exercises 1

1. The answer by simple reasoning is 3. The equation to prove this is $5 - 2 = 3$

2. The answer by simple reasoning is 8. The equation to prove this is $10 - 2 = 8$

3. He now has 6 kola nuts. The equation to prove this is $4 + 2 = 6$

4. He must buy 2 more t-shirts. The equation to prove this is $10 - 8 = 2$

5. 5 eggs will be left in the basket. The equation to prove this is $10 - 5 = 5$

Exercises 2

1.   (i) The unknown quantity is the number of pigs in the farm

  (ii) Let z = number of pigs in the farm

  (iii) Write the equation of the problem as

$$z = 54 + 70$$
$$z = 124$$

∴ there are 124 pigs in the farm

2.   (i) The unknown quantity is 'the number of toy soldiers that participated in the battle'

  (ii) Let z = the number of toy soldiers that participated in the battle

  (iii) Write the equation

$$z = 30 + 39 + 88 + 53$$
$$z = 210$$

∴ there are 210 toy soldiers that participated in the battle

3.   (i) The unknown is 'the age of Tom'

  (ii) Let z = age of Tom

  (iii) Write the equation

$$z = 18 - 12$$
$$z = 6$$

∴ Tom is 6 years old.

4. (i) The unknown is 'how many coins the man has'

   (ii) Let z = 'how many coins the man has'

   (iii) Write the equation

$$z = \text{(total coins)} - \text{(coins in the blue bag)}$$
$$z = (100 + 16 + 115) - 16$$
$$z = 231 - 16$$
$$z = 215$$

$\therefore$ The man has 215 coins

5. (i) The unknown is 'the number of fruits left in the tree'

   (ii) Let z = 'the number of fruits left in the tree'

   (iii) Write the equation

$$z = \text{(Initial fruits on tree)} - \text{(sum of fruits that fell)}$$
$$z = 50 - (6 + 13 + 30)$$
$$z = 50 - 49$$
$$z = 1$$

$\therefore$ only 1 fruit will be left on the tree

6. (i) The unknown is 'the number of birds alive'

   (ii) Let z = 'the number of birds alive'

   (iii) Write the equation

$$z = \text{(Initial number of birds)} - \text{(sum of birds that died)}$$
$$z = 300 - (15 + 11 + 20)$$
$$z = 300 - 46$$
$$z = 254$$

$\therefore$ 254 birds are alive

Exercises 3

1. (i) The unknown quantity is 'the number'

   (ii) Let z = 'the number'

   (iii) Write the equation

$$z = 1083 + 2102$$
$$z = 3185$$

$\therefore$ the number is 3185

2. (i) The unknown is 'the number of fruits'

(ii) Let z = 'the number of fruits'

(iii) Write the equation

$$z = 5 + 13$$
$$z = 18$$

∴ the number of fruits is 18

3. (i) The unknown quantity is 'the worth of a phone and a TV set'

(ii) Let z = the worth of a phone and a TV set

(iii) Write the equation

$$z = 12 + 200$$
$$z = \$212$$

∴ the worth of a phone and a TV set is \$212

4. (i) The unknown is the sum of 2299 and 1083

(ii) Let z = the sum of 2299 and 1083

(iii) Write the equation

$$z = 2299 + 1083$$
$$z = 3382$$

∴ the sum of 2299 and 1083 is 3382

5. (i) The unknown is 'the number of shoes which Martha has'

(ii) Let z = 'the number of shoes which Martha has'

(iii) Write the equation

$$z = 5 + 6$$
$$z = 11$$

∴ Martha has 11 pairs of shoes

Exercises 4

1. (i) The unknown is 'the result'

(ii) Let z = 'the result'

(iii) Write the equation

$$z = 5000 - 3085$$
$$z = 1915$$

∴ the result is 1915

2. (i) The unknown is 'the difference in price'

(ii) Let z = 'the difference in price'

(iii) Write the equation

$$z = 100 - 60$$
$$z = \$40$$

∴ the difference in price is \$40

3. (i) The unknown is the extra distance which Tom must walk

(ii) Let z = the extra distance which Tom must walk

(iii) Write the equation

$$z = 20 - 6$$
$$z = 14 \text{ km}$$

∴ the extra distance which Tom must walk is 14 km

4. (i) The unknown quantity is 'the number of trees lost'

(ii) Let z = 'the number of trees lost'

(iii) Write the equation

$$z = 2566 - 322$$
$$z = 2244$$

∴ the number of trees lost = 2244

5. (i) The unknown is the money in Jack's account minus the money he spent

(ii) Let z = the money in Jack's account minus the money he spent

(iii) Write the equation

$$z = 1000 - 60$$
$$z = \$940$$

∴ the money in Jack's account minus the money he spent = \$940

Exercises 5

1. (i) The unknown quantity is 'the number of items John gave Paul'

(ii) Let z = 'the number of items that John gave Paul'

(iii) Write the equation

$$z = (\text{initial number of pencils}) - (\text{final number of pencils})$$
$$z = (10 + 5 + 6) - (10)$$
$$z = 21 - 10$$
$$z = 11$$

∴ the number of items that John gave Paul = 11

2.   (i) The unknown is 'the money Mr. Jack had at first'

   (ii) Let z = 'the money Mr. Jack had at first'

   (iii) Write the equation

$$z = 500 + 2100$$
$$z = \$2600$$

$\therefore$ Mr. Jack had \$2600 at first

3. Part (a)

  (i) The unknown is the sum of the number of pencils owned by Sandra and James

  (ii) Let $z = \ldots$

Step 1: Let the number of pencils owned by Sandra be s and the number of pencils owned by James be j

$$\therefore z_a = s + j$$

Step 2:

$$j = s - 6$$
$$s = 22 + 5 = 27$$
$$\therefore j = 27 - 6$$
$$j = 21$$

$$\therefore z_a = 27 + 21$$
$$z_a = 48$$

Part (b)

  (i) The unknown is the sum of the number of pencils owned by Lily and James

  (ii) Let $z_b = \ldots$

    Let l = number of pencils owned by Lily

  (iii) Write the equation

$$z_b = l + j \quad (\text{recall } j = 21 \text{ from part(a)})$$
$$z_b = 22 + 21$$
$$z_b = 43$$

Part (c)

  (i) The unknown is the total number of pencils

  (ii) Let z_c = total number of pencils

(iii) Write the equation
$$z_c = l + s + j + b$$
where b = number of pencils owned by Ben. Since $l = 22$, $s = 27$, $j = 21$, $b = 31$

$$z_c = 22 + 27 + 21 + 31$$
$$z_c = 101$$

$\therefore$ the total number of pencils is 101

4. (i) the unknown is the number of girls present in the class

   (ii) Let z = number of girls present in the class

   (iii) Write the equation

$$z = 33 - 13$$
$$z = 20$$

$\therefore$ there are 20 girls present in the class

5. (i) the unknown is 'the number of white cards'

   (ii) Let z = 'number of white cards'

   (iii) Write the equation

$$z = (\text{total number of cards}) - (\text{green} + \text{yellow} + \text{black} + \text{red})$$

From the question we are given
total number of cards = 40
number of green cards = 5, number of yellow cards = 10, number of black cards = 15, number of red cards = 2.
Therefore

$$z = 40 - (5 + 10 + 15 + 2)$$
$$z = 40 - 32$$
$$z = 8$$

$\therefore$ there are 8 white cards

Exercises 6

1. (i) the unknown is the age of Baba

   (ii) Let z = the age of Baba

   (iii) Write the equation
$$z = 10 + y$$

$\therefore$ Baba will be $10 + y$ years old in y years.

2. (i) the unknown is 'the number of apps altogether'

   (ii) Let z = the number of apps altogether

(iii) Write the equation

$$z = 12 + y$$

∴ there are $12 + y$ apps altogether

3. (i) the unknown is 'Richard's mark'

(ii) Let $z =$ Richard's mark

(iii) Write the equation

$$z = x + 20 = 20 + x$$

∴ Richard's mark is $20 + x$

4. (i) the unknown is 'number of fruits the boy has'

(ii) Let $z =$ 'the number of fruits the boy has'

(iii) Write the equation

$$z = a + b$$

∴ the boy has $a + b$ fruits altogether

5. (i) the unknown is 'her age y years ago'

(ii) Let $z =$ 'her age y years ago'

(iii) Write the equation

$$z = 30 - y$$

∴ the woman was $30 - y$ years old y years ago

6. (i) the unknown is 'the number of pages which the boy has not read'

(ii) Let $z =$ 'number of pages which the boy has not read'

(iii) Write the equation

$$z = w - 30$$

∴ the number of pages which the boy has not read is $w - 30$

7. (i) the unknown quantity is 'the length of the line'

(ii) Let $z =$ 'the length of the line'

(iii) Write the equation

$$z = 15 - x$$

∴ the length of the line is "$15 - x$"

8. (i) the unknown quantity is 'the profit'

(ii) Let $z =$ 'the profit'

(iii) Write the equation

$$\text{Net profit} = (\text{Selling price}) - (\text{Cost price})$$
$$\therefore z = (\text{Selling price}) - (\text{Cost price})$$

In the question, selling price $= \$165,500$, cost price $= \$x$

$$\therefore z = 165,500 - x$$

∴ the profit is $165,500 - x$

9. (i) the unknown quantity is 'the distance he has to travel'

(ii) Let $z = $ 'the distance he has to travel'

(iii) Write the equation

$$z = 20 - x$$

$\therefore$ the distance he has to travel is $20 - x$ km

10. (i) the unknown quantity is 'the money left in his savings'

(ii) Let $z = $ 'the money left in his savings'

(iii) Write the equation

$$z = s - 100$$

$\therefore$ 'the money left in his savings is '$s - 100$' dollars

11. (i) the unknown quantity is 'the number of animals the farmer has'

(ii) Let $z = $ the number of animals the farmer has

(iii) Write the equation

$$z = 6 + m + n + 5$$
$$z = 6 + 5 + m + n$$
$$z = 11 + m + n$$

$\therefore$ the farmer has $11 + m + n$ animals in his farm

12. (i) the unknown quantity is 'the number of fruits left at the end of the day'

(ii) Let $z = $ the number of fruits left at the end of the day

(iii) Write the equation

$$z = 12 - m - n - 6$$
$$z = 12 - 6 - m - n$$
$$z = 6 - m - n$$

$\therefore$ the number of fruits left at the end of the day is $6 - m - n$

13. (i) the unknown quantity is 'the number of pencils left in the box'

(ii) Let $z = $ 'the number of pencils left in the box'

(iii) Write the equation

$$z = k - 5 + 25 - h$$
$$z = k + 20 - h$$

$\therefore$ there are $k + 20 - h$ pencils left in the box

14. (i) the unknown is the number of pencils they have altogether

(ii) Let $z = $ 'the number of pencils they have altogether'

(iii) Write the equation

$$\text{Let the number of Sandra's pencils} = s$$
$$\text{the number of James pencils} = j$$
$$\text{the number of Lily's pencils} = n$$

1. $s = n + 5$

2.

$$j = s - p$$
$$\therefore j = n + 5 - p$$

3.

$$z = (\text{Lily}) + (\text{Sandra}) + (\text{James})$$
$$z = n + s + j$$
$$z = n + (n + 5) + (n + 5 - p)$$
$$z = n + n + 5 + n + 5 - p$$
$$z = n + n + n - p + 5 + 5$$
$$z = 3n - p + 10$$

$\therefore$ the number of pencils that have altogether is $3n - p + 10$

15. (i) the unknown is 'the number of blue cards'

(ii) Let $z =$ 'the number of blue cards'

(iii) Write the equation

$$z = 40 - t - 6$$
$$z = 40 - 6 - t$$
$$z = 34 - t$$

$\therefore$ the number of blue cards is $34 - t$

Exercises 7

1. (i) the unknown is 'the number of days in five weeks'

(ii) Let $z =$ 'the number of days in five weeks'

(iii) Write the equation

$$z = 7 \times 5$$
$$z = 35$$

$\therefore$ there are 35 days in five weeks

2. (i) the unknown is 'the number of days in fifteen years'

(ii) Let $z =$ 'the number of days in fifteen years'

(iii) Write the equation

$$z = 366 \times 15$$
$$z = 5490$$

∴ there are 5490 days in fifteen years

3.   (i) the unknown is 'the cost of 5 books'

  (ii) Let $z =$ the cost of 5 books

  (iii) Write the equation

$$z = 5 \times 30$$
$$z = \$150$$

∴ the cost of 5 books is \$150

4.   (i) the unknown is 'the number 9 times greater than 5'

  (ii) Let $z =$ 'the number 9 times greater than 5'

  (iii) Write the equation

$$z = 9 \times 5$$
$$z = 45$$

∴ the number 9 times greater than 5 is 45

5.   (i) the unknown is 'how much he would have earned'

  (ii) Let $z =$ 'how much he would have earned'

  (iii) Write the equation

$$z = 8 \times 8$$
$$z = \$64$$

∴ he would have earned \$64

6.   (i) the unknown is 'the number of chocolate bars Bisi has in total'

  (ii) Let $z =$ 'the number of chocolate bars Bisi has in total'

  (iii) Write the equation

$$z = 10 \times 16$$
$$z = 160$$

∴ Bisi has 160 chocolate barks in total

Exercises 8

1. (i) the unknown quantity is 'the number of sheets of paper she has left'

   (ii) Let z = 'the number of sheets of paper she has left'

   (iii) Write the equation

   (1) the number of sheets she gave the students is

   $$5 \times 62 = 310$$

   (2) the number of sheets she has left is z,

   z = (the initial number of sheets) − (the number of sheets she gave the students)

   $z = 400 - 310$

   $z = 90$

   ∴ she has 90 sheets of paper left

2. **Part (a)**

   (i) the unknown is 'the number of hot dogs which Josie ate'

   (ii) Let z_a = the number of hot dogs which Josie ate

   (iii) Write the equation

   $$z_a = 27 \times 7$$
   $$z_a = 189$$

   ∴ Josie ate 189 hot dogs

   Part (b)

   (i) the unknown is 'how many more hot dogs Josie ate than Sally'

   (ii) Let z_b = how many more hot dogs Josie ate than Sally

   (iii) Write the equation

   z_b = (number of hot dogs Josie ate) − (number of hot dogs Sally ate)

   $z_b = 189 - 27$

   $z_b = 162$

   ∴ Josie ate 162 more hot dogs than Sally

3. (i) The unknown quantity is 'the number of kilometers Bethel ran'

   (ii) Let z = 'the number of kilometers Bethel ran'

   (iii) Write the equation

   $$z = (\text{distance Jude ran}) - (15)$$

   distance Jude ran $= 4 \times 9 = 36$km

   $$\therefore z = 36 - 15$$
   $$z = 21\text{km}$$

   ∴ 'the number of kilometers Bethel ran is 21.'

4. (i) The unknown quantity is 'the difference'

    (ii) Let z = 'the difference'

    (iii) Write the equation.

$$z = (100) - \text{(a number that is nine times greater than 6)}$$

$$\text{(a number that is nine times greater than 6)} = 9 \times 6 = 54$$

$$\therefore z = 100 - 54$$
$$z = 46$$

$\therefore$ 'the difference is 46'

5. (i) The unknown quantity is 'the number of chocolate boxes which Bisi has'

    (ii) Let z = 'the number of chocolate boxes which Bisi has'

    (iii) Write the equation

$$z = \text{(number of chocolate boxes)} \times 100$$

number of chocolate boxes = (initial number of chocolate boxes) − (number of chocolate boxes sold)

$$= 16 - 5$$
$$= 11$$
$$\therefore z = 11 \times 100$$
$$z = 1100$$

$\therefore$ the number of chocolate bars which Bisi has is 1100.

Exercises 9

1. (i) The unknown quantity is 'the money lost on the car'

    (ii) Let z = 'the money lost on the car'

    (iii) Write the equation

$$z = \text{(cost price)} - \text{(selling Price)}$$
$$z = p - \frac{4}{5} \times p$$
$$z = \frac{p}{1} - \frac{4p}{5} = \frac{5p - 4p}{1 \times 5} = \frac{p}{5}$$
$$\therefore z = \frac{p}{5}$$

2. (i) The unknown quantity is 'the age of the father in y years'

    (ii) Let z = 'the age of the father in y years'

(iii) Write the equation

$$z = (\text{Present age of the father}) + y$$

$$(\text{Present age of father}) = 5 \times x = 5x$$

$$\therefore z = 5x + y$$

3. (i) The unknown quantity is 'the number of cattle the man has'

    (ii) Let z = 'the number of cattle the man has'

    (iii) Write the equation

$$z = n \times p$$

$$z = np$$

$\therefore$ The man has np cattle ranches.

4. (i) The unknown quantity is the number of days in t years.

    (ii) Let z = number of days in t years

    (iii) Write the equation

$$z = 366 \times t$$

$$z = 366t$$

5. (i) The unknown quantity is 'the cost of the coat and pair of trousers'

    (ii) Let z = the cost of the coat and pair of trousers

    (iii) Write the equation

$$z = (\text{cost of coat}) + (\text{cost of pair of trousers})$$

$$\text{cost of pair of trousers} = 100$$
$$\text{cost of coat} = p \times 100 = 100p$$
$$\therefore z = \$(100p + 100)$$

$\therefore$ the cost is $\quad \$(100p + 100)$

Exercises 10

1. **Part (a)**

    (i) The unknown quantity is 'the number of weeks'

    (ii) Let z = 'the number of weeks'

    (iii) Write the equation.

$$z = \frac{84}{7}$$

(We use 7 in the denominator because there are 7 days in one week) $\therefore z = 12$

$\therefore$ the number of weeks is 12

Part b

(i) The unknown quantity is 'the number of weeks'

(ii) Let z = 'the number of weeks'

(iii) Write the equation

$$z = \frac{21}{7}$$
$$\therefore z = 3$$

$\therefore$ the number of weeks is 3.

2. (i) The unknown quantity is 'the money each classmate gave to Richard'

(ii) Let z = 'the money each classmate gave to Richard

(iii) Write the equation

$$z = \frac{10\cancel{0}}{2\cancel{0}} = \frac{10}{2}$$
$$z = 5 \quad \text{dollars}$$

$\therefore$ the money each classmate gave to Richard is \$5

3. (i) The unknown quantity is 'the number of years'

(ii) Let z = 'the number of years'

(iii) Write the equation.

$$z = \frac{732}{366}$$

(We use 366 in the denominator because there are 366 days in one year)

$$\therefore z = 2$$

$\therefore$ the number of years is 2.

Part a

4. (i) The unknown quantity is 'the unit price'

(ii) Let z = 'the unit price'

(Note: unit price refers to the price of one item, or in this case the unit price means the price of one book).

(iii) Write the equation.

$$z = \frac{3200}{8}$$
$$z = 400 \quad \text{dollars}$$

$\therefore$ the unit price is \$400

Part b

(i) The unknown quantity is 'the unit price'

(ii) Let z = 'the unit price'

(iii) Write the equation.

$$z = \frac{150}{5}$$
$$\therefore z = 30 \quad \text{dollars}$$

$\therefore$ the unit price is \$30

5. (i) The unknown quantity is the 'the number'

(ii) Let z = 'the number'

(iii) Write the equation.

$$z = \frac{1}{3} \times 9$$
$$z = \frac{9}{3}$$
$$z = 3$$

$\therefore$ the number is 3.

Exercises 11

1. (i) The unknown quantity is 'the amount of drink in each glass'

(ii) Let z = 'the amount of drink in each glass'

(iii) Write the equation

$$z = \frac{(\text{total amount of juice})}{2}$$

The denominator is 2 because there are two glasses.

$$(\text{total amount of juice})$$
$$= 50\text{ml} + 30\text{ml}$$
$$= 70\text{ml}$$
$$\therefore z = \frac{70}{2}\text{ml}$$
$$z = 35\text{ml}$$

$\therefore$ the amount of drink in each glass is 35ml

2. (i) The unknown quantity is 'the length of ribbon each person gets'

(ii) Let z = 'the length of ribbon each person gets'

(iii) Write the equation.

$$z = \frac{\text{(total length of ribbon)}}{3}$$

We use 3 in the denominator because the ribbon is shared between 3 persons (daughters)

$$\text{(total length of ribbon)} = 42\text{cm} + 24\text{cm}$$
$$= 66\text{cm}$$
$$\therefore z = \frac{66\text{cm}}{2}$$
$$z = 33cm$$

$\therefore$ the length of ribbon each person gets is 33cm

3.   (i) The unknown quantity is 'the number if tables needed'

(ii) Let $z = $ 'the number of tables needed'

(iii) Write the equation

$$z = \frac{\text{total number of people}}{5}$$

The denominator is 5 because each table can seat 5 people

$$\text{(total number of people)} = 25 + 5$$
$$= 30 \quad \text{people}$$
$$\therefore z = \frac{30}{5}$$
$$z = 6$$

$\therefore$ 6 tables are needed
Or the number of tables needed is six.

4.   (i) The unknown quantity is 'the number of chickens in each group'

(ii) Let $z = $ 'the number of chickens in each group'

(iii) Write the equation.

$$z = \frac{\text{(remaining chickens)}}{4}$$

The denominator is 4 because the remaining chickens were kept in 4 equal groups

$$\text{(remaining chickens)}$$
$$= \text{(initial number of chickens)} - \text{(number of chickens sold)}$$
$$= 35 - 3$$
$$= 32$$
$$\therefore z = \frac{32}{4}$$
$$z = 8$$

$\therefore$ there are 8 chickens in each group

5. (i) the unknown quantity is 'the money each child gets'

(ii) Let z = 'the money each child gets'

(iii) Write the equation

$$z = \frac{\text{(total money won by Mrs Jones)}}{3}$$

We use '3' because Mrs Jones shared the money between her 3 children. But

$$\text{(total money won by Mrs Jones)} = €45 + €30$$
$$= €75$$
$$\therefore z = \frac{75}{3}$$
$$z = 25$$

∴ the money each child gets is €25

Exercises 12

1. (i) The unknown quantity is 'the number of children on each coach'

(ii) Let z = 'the number of children on each coach'

(iii) Write the equation.

$$z = \frac{\text{total number of children}}{p}$$

We use 'p' because there are p coaches.

$$\text{(total number of children)} = 120 + 130$$
$$= 250 \quad \text{children}$$
$$\therefore z = \frac{250}{p}$$

∴ the number of children on each coach is $\frac{250}{p}$

2. (i) The unknown quantity is 'the amount of drink in each glass'

(ii) Let z = 'the amount of drink in each glass'

(iii) Write the equation

$$z = \frac{\text{total amount of drink}}{8}$$

We use 8 because there are 8 glasses.

$$\text{(total amount of drink)} = x + 680$$
$$\therefore z = \frac{x + 680}{8}$$

∴ the amount of drink in each glass is $\frac{x+680}{8}$ ml

3. (i) The unknown quantity is 'the number of bricks in each box'

(ii) Let z = 'the number of bricks in each box'

(iii) Write the equation

$z = \frac{\text{(total number of bricks)}}{p}$ The denominator is p because the bricks were split equally into p boxes. but

$$(present\,number\,of\,bricks)$$
$$= \text{(initial number of bricks)} - \text{(number of bricks sold)}$$
$$= m - n$$
$$\therefore z = \frac{m - n}{p}$$

$\therefore$ the number of bricks in each box is $\frac{m-n}{p}$

4. (i) The unknown is 'the number of rows'

   (ii) Let z = 'the number of rows'

   (iii) Write the equation

   $z = \frac{\text{(total number of people)}}{m}$ The denominator is m because each row can seat m people or there are m people in a row

   $$\text{(total number of people)} = \text{(People in stall)} + \text{(People in balcony)}$$
   $$= 360 + t$$
   $$\therefore z = \frac{360 + t}{m}$$

   $\therefore$ there are $\frac{360+t}{m}$ rows

5. (i) The unknown quantity is 'the money each child gets'

   (ii) Let z = 'the money each child gets'

   (iii) Write the equation.

   $$z = \frac{\text{total money won}}{n}$$

   The denominator is n because the money was shared among n children.

   $$\text{but (total money won)} = x + 1030$$
   $$\therefore z = \frac{x + 1030}{n}$$

   $\therefore$ each child gets $\left(\frac{x+1030}{n}\right)$

Exercises 13

Part a

1. (i) The unknown quantity is 'the number of cookies in each pile'

   (ii) Let z = 'the number of cookies in each pile'

(iii) Write the equation

$$z = \frac{18}{5} = 3\frac{2}{5}$$

Since z must be a whole number, take only the whole number as the answer.

$\therefore z = 3$

$\therefore$ the number of cookies in each pile is 3

Part b

(i) The unknown quantity is 'the number of leftover cookies'

(ii) Let z_b = 'the number of left-over cookies'

(iii) To get z_b, multiply z_a from **Part (a)** with 5 and subtract the result from 18

$$\text{i.e} \quad z_b = 18 - z_a \times 5$$
$$z_b = 18 - 15$$
$$z_b = 3$$

$\therefore$ the number of left-over cookies is 3

2. (i) The unknown quantity is 'the money Ben got'

(ii) Let z = 'the money Ben got'

(iii) Write the equation

Step 1: $\frac{50}{3} = 16\frac{2}{3}$

the whole number is 16

step 2:

$$z = 50 - (16 \times 3)$$
$$z = 50 - 48$$
$$z = 2$$

$\therefore$ the money Ben got is \$2

3. (i) The unknown quantity is 'the money each child gets'

(ii) Let z = 'the money each child gets'

(iii) Write the equation

Step 1: total money Mrs Peters won $= 4500 + 3100 = \$7600$

Step 2:

$$z = \frac{\text{total money won by Mrs Peters}}{3}$$
$$z = \frac{7600}{3}$$
$$z = 2533\frac{1}{3}$$

take the whole number of the mixed fraction as the answer

i.e $z = 2533$

$\therefore$ the money each child gets is \$2533

4. (i) The unknown quantity is 'the number of chickens Micheal kept'

(ii) Let $z =$ 'the number of chickens Micheal kept'

(iii) **Step 1:**

$$\frac{\text{remaining chickens}}{4} = \frac{35 - 4}{4} = \frac{31}{4}$$
$$= 7\frac{3}{4}$$

Step 2: the whole number is 7

$$z = 31 - (4 \times 7)$$
$$z = 31 - 28$$
$$z = 3$$

5. (i) The unknown quantity is 'the money he made'

(ii) Let $z =$ 'the money he made'

(iii) Write the equation

$$z = (\text{the number of remainder sweets}) \times \$10$$

Step 1:

(i) Total number of sweets $= 46 + 31 = 77$

(ii) $\frac{\text{total number of sweets}}{5 \text{ friends}} = \frac{77}{5} = 15.4$

Step 2:   The whole number is 15

$$\therefore (\text{remainder}) = 77 - (15 \times 5) = 77 - 75 = 2$$

$$z = (\text{remainder}) \times 10 = \$20 \quad \therefore \text{ Joshua made } \$20$$

Made in the USA
Columbia, SC
29 May 2025

58604177R00059